Dog Gone Missing

Why Dogs Go Missing and How to Find Them

Amy Mall

Oh Street Press
Dog Gone Missing:
Why Dogs Go Missing and How to Find Them
Amy Mall

Copyright ©2017 by Amy Mall

All Rights Reserved.

This book was self-published by the author, Amy Mall, and Oh Street Press. No part of this book may be reproduced in any form by any means without the express permission of the author. This includes reprints, excerpts, photocopying, recording, or any future means of reproducing text.

Published in the United States of America by Oh Street Press in 2017.

ISBN 978-0-692-98042-2

Book cover: Lindsay Peternell and Bruce Marsh
Interior design: Lindsay Peternell

www.doggonemissing.com

Disclaimer: This book is designed to provide helpful information that is accurate to the best of our knowledge. Any recommendations are made for informational purposes only, without any guarantee on the part of the author, and are not meant to be an endorsement of any service or product. The author and publisher disclaim any liability in connection with the use of the information in this book.

NOTE TO OWNERS OF MISSING DOGS

If your dog is lost right now and you are in need of immediate advice, skip to Chapter 10 for the top ten ways to find your missing dog.

Come back for the rest later.

To my dogs:

Companions, friends, guardians, teachers, and troublemakers.

If only we knew what our dogs were really thinking...

Contents

Introduction	1
1. The Dog-Human Bond	7
2. Is My Dog Like Lassie?	13
3. How Wolves Navigate	22
4. The Science of Dogs and Navigation	27
5. Scout	32
6. Why Dogs Go Missing	40
7. Stolen Dogs	45
8. Dora	53
9. Top Ten Tips to Keep Your Dog Home	58
10. Top Ten Tips to Find Your Missing Dog	74
Epilogue: Grieving a Missing Dog	106
Acknowledgements	109
Appendix: Online Resources	111
Notes	113

Introduction

It's a phone call that no dog owner wants to get.

I'll never forget that day during the summer of 2001. I was living on Capitol Hill in Washington, D.C., a charming historic neighborhood with row houses, small parks, and a lively street life. My best friend and companion was Lola, three years old at the time, a German shepherd/Samoyed mix with soft, fluffy fur the color of ivory.

I was working downtown, a few miles away, when one afternoon I received the dreaded call at my office. It was from a contractor my landlord hired for a home renovation project. "Your dog ran away," I heard him say.

My heart dropped. What? How? Why? And most important: where was she? I was stunned; I always felt that Lola and I had a tight bond. She'd never shown any interest in running away from home. We'd driven across the country and gone many places together. When we hiked in the mountains, she was reliable off-leash and liked to take the lead—but was never too far ahead and would constantly look back at me. I always told myself that Lola walked ahead to make sure our path was safe, and looked back to check that I was still there. She was protective by nature and not the type to "go walkabout" alone.

Lola, *photo by Amy Mall*

The contractor and his crew had been working in the house for weeks, and Lola seemed comfortable with them. She gave no indication that she was afraid or wanted to get away from them for any other reason. Nevertheless, she had run.

The contractor said Lola bolted when one of the workers opened the front door. He told me that they had seen her run down the steps and turn left, headed south. That news was more cause for alarm: a few blocks south she would

Introduction

encounter Pennsylvania Avenue, a major thoroughfare, just as rush hour was approaching.

Worried sick, I left my office to get home. Lola was the first dog I'd adopted, and I didn't know anything about searching for a lost pup. Everyone I knew was still at work, so they couldn't help. Those were the days before I automatically Googled every question that popped into my head, so it didn't occur to me to search for guidance on the internet. When I got home, I hopped in my car, thinking I could cover more ground that way. I started driving slowly, headed south. I scanned the sidewalks on each side of the street, calling for Lola, hoping with all my heart she hadn't gotten hit by a car, and praying I'd find her soon.

After about two hours of searching with no sign of Lola, my sense of panic increased. Then my cell phone rang. My friend Melanie lived only a few blocks away from me with her shepherd mix Cheyenne, Lola's best buddy. Now she was on the other end of the phone. Melanie had just gotten home from work. "Why is Lola lying behind a bush in my front yard?" she asked.

I let out a deep sigh with enormous relief. Apparently, Lola had been comfortably chilling in her pal's front yard until Melanie got home. All had ended well after a relatively short period of time.

Once I calmed down, I started wondering about Lola's escape. Many questions rushed through my mind. Maybe she ran out the door because she was afraid and, since I wasn't home, decided that Melanie's house was a safe place. On the other hand, she might have seen the open door as an opportunity for fun, and decided to arrange her own playdate with her canine friend Cheyenne. I wondered whether she had indeed run south, as I'd been told, and then turned around at some point, or whether the workers had somehow gotten it wrong. Lola's adventure was a mystery.

> **Lost dog lesson #1:**
> Take extra precautions when other people will be in your house and might open a door. Your dog might slip out the door and run, even if they wouldn't do that when you are at home.

I was lucky; this was a brief crisis that had a happy ending. Fortunately, many worried owners quickly find their lost dogs. But not every missing dog story ends happily ever after. It's devastating when dogs run away and are never found, their owners left with broken hearts and gut-wrenching fear about what happened and whether they will ever see their dog again.

Over the years, I've continued to wonder what goes on inside a dog's mind when they run away. Whether it's a dog I've read about in an article, or seen on a poster, I've wanted to know why dogs run away, where they go, if they are lost, or if they choose not to return home. More than just simple curiosity, the answers to these questions could provide truly valuable insight for dog owners regarding how we can prevent our dogs from running away and how best to find a lost dog.

Many people must be asking the same questions, because a lot of us share our lives with dogs. In 2015, the American

Introduction

Pet Products Association estimated that almost half of American households—more than 54 million—owned at least one dog. Altogether the organization estimated that there are about 78 million pet dogs in the United States.

Many of these family pets go missing every year. Experts with the American Society for the Prevention of Cruelty to Animals (ASPCA) conducted a random survey of dog owners and found that 14 percent of dogs were lost at least once during a five-year period. Seven percent of the missing dogs "were never reunited with their owners."

Seven percent may sound small, but if there are approximately 78 million dogs in the United States, it amounts to about 5.5 million dogs. In other words, based on these surveys, an average of more than one million dogs might go missing each year, and the number could be even greater. According to the American Kennel Club, "One out of every three pets will go missing at some point in their lifetime."

This book is for you if you've ever wondered about what goes on in your dog's mind, why a family dog might run away, and why some dogs can find their way home while others can't. We'll explore whether the essence of dogness is a dog's bond to their human, or a desire to explore the world outside their home.

As you read through the book, you'll find ten "lost dog lessons." If you've ever tried to find a lost dog, or worried that your dog might run away, Chapters 1 through 4 explore the science behind the dog-human emotional bond, whether dogs can navigate like their close wolf relatives, and how a dog's intelligence might help them find their way home. We'll look at what the research and experts say about the canine homing instinct and sense of direction, as well as personal stories about dogs and their skills.

Chapters 5 through 8 investigate why dogs go missing from their human families and where they might go, including real-life stories of lost and stolen dogs and how they were found. The final chapters, Chapters 9 and 10, discuss proven steps to keep your dog safely at home. You'll find practical recommendations you can employ right now to help reduce the chance that your dog will run away or get lost in the future. I've also included key steps you can take to be prepared for the worst if it ever happens, and increase the chances you will find your dog, and find them quicker, if they do go missing. Most important for some readers, you'll find the top ten most effective ways to search for a beloved canine who is missing right now.

I was lucky that day when Lola ran out the front door and into the streets of Washington by herself. She didn't go far, made it safely to her destination, and was found within a few hours. But having a dog go missing, especially one who isn't found soon, can be a distressing combination of terror and heartbreak. This book can help. What follows is not only science, but also stories of inspiration and hope—stories that illustrate what both our dogs and we are capable of if we ever become separated from each other.

1

The Dog-Human Bond

Almost half of us in the United States share our lives with dogs. Our dogs are our best friends and are considered family members. Some of us even call our dogs our "fur babies." Pampered pups get their own clothing, furniture, toys, massages, acupuncture, and more. We talk to our dogs. We miss them and worry about them when we travel away from home. At the end of their lives, we deeply mourn their deaths.

There is no doubt that we humans love our dog companions and feel a powerful connection to them. Most of us believe that our dogs return our love and affection, feeling the same strong emotional attachment. Fortunately, science provides some insight into what our dogs really feel about us.

Origins of the Bond

People have been curious, even obsessed, about the evolutionary origins of the dog for a very long time. There are researchers around the world who are dedicated to investigating where, when, and how dogs became the domestic dog that we know today—man's best friend. Every time new fossils and pieces of archeological evidence are discovered in different corners of the globe, new theories are proposed, analyzed, dissected, and discussed. Even with the latest DNA technology, however, there are still missing pieces in this complicated scientific puzzle.

To this day, scholars continue to engage in what is sometimes heated debate about the ancestry of our dogs. While a consensus is lacking on the exact moment, location, or process of the dog's evolution, there is broad agreement on at least one thing: dogs descended from a wolf species, an ancestor of the modern-day wolf. It has been reported that dogs and modern wolves share 99.96 percent of their DNA.

Thousands of years ago, wolves had many things in common with humans of the time. They often shared territory. Both depended on a social structure for survival—a pack for wolves, the family for humans. They both hunted in groups and hunted for some of the same prey, enjoying some of the same food. Both excelled at problem solving, communicated by vocalizing, and were friendly while at the same time protective.

Wolves and humans also had their differences. Wolves, for example, were better at chasing prey with their speed and agility, and their superior hearing allowed them to better detect possible threats. Humans, however, had tools and weapons that made them more successful at numerous tasks. This unique mix of commonalities and different strengths eventually led to these two species forming a relationship and living symbiotically alongside each other in prehistoric times, before there ever was such a thing as a domestic dog.

Mark Derr, author of *How the Dog Became the Dog*, calls wolves that took up with humans "socialized wolves." He writes that, absent evidence to the contrary, ".... the association between socialized wolves and humans was consensual and mutual, and in response to the needs and desires of both species, as well as to exigencies of rapidly shifting environmental conditions. They helped each other out, and they adapted together to a changing world." Eventually, some of these wolves came to be what we now know as pets, and the domestic dog was born.

It's fascinating to think that, before dogs existed, man's best friend might have been the wolf. While everyone recognizes that the domestic dog descended from the wolf, experts don't always agree about which traits dogs have in common with their wolf cousins. Sharing so much genetic material doesn't mean they are exactly alike. Humans share approximately 99 percent of our DNA with bonobos and chimpanzees, yet we are quite different from those species.

The Science of the Bond

The ancient ties between humans and wolves seem to have established a firm foundation for the powerful bonds that were yet to come between humans and modern dogs. Dog lovers have long felt a strong connection with the canine members of our families. It's not only our instincts telling us that a special dog-human connection exists. Scientists have confirmed that there is an actual chemical bond between humans and dogs.

When humans and their dogs gaze into each other's eyes, both begin to experience higher levels of the chemical oxytocin. Oxytocin is a chemical that has been called the "love hormone," and even the "cuddle hormone," because it promotes bonding. This oxytocin increase in both beings is what's known as a "positive-feedback loop," where an increase in one leads to an increase in the other. The same loop has been documented between human parents and their babies.

In addition to measuring chemical reactions between dogs and their owners, researchers around the world have conducted experiments to see how dogs physically react to their owners, and to compare that to how they react to strangers. At Emory University, researchers used MRI scanners to monitor how the brains of dogs respond to different scents. They discovered that dogs' brains react very differently to the sweat of a familiar human than they do to the scent of strangers, or even to the scent of familiar dogs.

Hungarian scientists observed shelter dogs who were placed in unfamiliar situations, either with people they'd met before or with strangers. The dogs showed a clear preference for familiar humans by interacting with them more and following them. The researchers concluded that the shelter dogs were drawn to social interaction with humans and formed a meaningful bond in a short period of time.

In another experiment, puppies up to eight weeks old were isolated in order to induce "separation-induced distress vocalization" or, in non-scientific terms, barking or crying. Researchers studied the response of the isolated puppies to 12 different stimuli, including food, toys, other dogs, and humans. Out of the 12 different situations, contact with humans led to the greatest decrease in vocalization among the puppies. In other words, their stress was lowest when they were around humans, even when compared to being around other dogs.

Scientists at the Clever Dog Lab at the University of Veterinary Medicine in Vienna, Austria, conducted experiments in which they gave dogs various puzzle toys with food inside, and the dogs had to figure out how to get the food out. They found that dogs would spend much more time trying to solve the puzzle when their owner was in the room. This was the case even after controlling for various factors, such as the owner's behavior when they were in the room with their dog, the dog's interest in food, and the dog's stress level. The researchers concluded that both young and adult dogs feel more secure when their owners are nearby.

Dogs vs. Wolves: Pack Bonds

The powerful attachment of dogs to their humans makes them beloved pets, and also helps them excel at certain specialized jobs. It turns out that the ability of dogs to bond with humans makes them a true stand-out in the animal kingdom. Wolves are social and can be very friendly with humans but, even when they are raised by people, they don't experience the oxytocin positive-feedback loop or develop the kind of chemical bonds to their humans that dogs do.

The powerful dog-human bond is one reason dogs are often referred to as "pack animals," a term that popular culture has borrowed from wolves. Yet despite these intense emotional attachments, dogs still run away from their human families. Often. It turns out that dogs aren't pack animals in the way that wolves are, or in the way we might think.

Marc Bekoff, professor emeritus of ecology and evolutionary biology at the University of Colorado, told me that the average dog who has been treated well and raised in a healthy environment likes to feel that they are part of a group. Being a member of a human family, he said, is a meaningful association for a dog. But a human family isn't the same as a wolf pack. It doesn't operate or determine survival in the same way. Ultimately, it doesn't hold the same significance for a dog that a pack holds for its wolf relatives.

Anneke Lisberg, associate professor of biological sciences at the University of Wisconsin, told me that, while dogs form very strong social bonds, they also live "in the moment." This can explain why a dog may get separated from their family. "They are potentially following an animal or something exciting, or maybe they're lost and confused, or they're just so freaked out by a car that just went by that they're just bolting and panicking. It can be any number of reasons that they go, but it's not because they don't like their owner. It's that, in the moment, there is something else happening."

Lost dog lesson #2:
Don't assume that your dog will decide to come home on their own because of pack bonds. Dogs don't feel pack bonds the way that wolves do.

2

Is My Dog Like Lassie?

Lassie is probably the most famous dog of all time. A beautiful collie who appeared in books, films, and television programs, she had renowned navigation skills and could find her way home over long distances through unfamiliar and rough terrain despite an ongoing series of seemingly insurmountable challenges. Readers and viewers were fascinated with her for generations. Then again, she was fictional.

But beyond fictional accounts, we've all seen news articles and video clips about real dogs with miraculous tales of navigation and survival, some finding their way home after years away, some over distances of hundreds (or even thousands!) of miles.

Since that day in 2001 when Lola ran away and found her way to Melanie's house, I've shared my life with three other dogs: Louie, Roxanne, and Shanti. All were mixed-breed rescues of various origins, found as strays. As dog owners know, every dog has their own personality and talents, and I loved each of my dogs for their quirky individuality. My relationship with each one was a little different, and from each I learned distinct lessons about a dog's attachment to their family, as well as the canine homing instinct.

Roxy: Fear and Hiding

Roxanne (a.k.a. Roxy) was a red-headed mix of unknown breed and origin who came to my home as a foster dog. When first rescued, she was terrified of humans. A trainer thought she was "semi-feral." At first, Roxy wouldn't eat or drink if a human was in the same room. She wouldn't make eye contact. On walks she would try to crawl under parked cars. She took every opportunity she could to hide behind or beneath something. I sometimes found her peeking out from behind the headboard of my bed, or under the back porch. As much as I prided myself on being a dog person, I had to admit that she seemed to feel no attachment to me at all. I tried not to take it personally.

Although she was terrified around people or loud noises, I saw glimpses of unrestrained playfulness in Roxy. When alone, she would joyfully run in circles around the yard, throw a toy up in the air and try to catch it. She also got along well with Lola, my other dog. Since it seemed unlikely that anyone else would ever adopt her, given her disinterest in humans, she ultimately became what is known in the dog rescue world as a "failed foster." In other words, her foster mom adopted her! She became a full-fledged member of the family.

Is My Dog Like Lassie?

Roxanne, *photo by Amy Mall*

Shortly after I adopted Roxy, we moved to Colorado. One summer day in the Sangre de Cristo Mountains, I was cleaning a cabin where we had been staying, nine miles from the nearest town. Roxy got spooked by the sound of the vacuum cleaner and ran out an open door. By the time I realized she was missing, there was no sign of her, and she didn't come when called. My then-boyfriend and I began looking for her. We called her name for hours and, assuming she had kept running once she got out the door, gradually expanded our search farther and farther from the cabin.

After a few hours of fruitless searching, a memory popped into my mind of the last time I was searching for Roxy, months earlier, in Washington, D.C., when she had gotten away from us in a park. We had eventually found her when someone spotted the end of her leash sticking out from under a bush. A light bulb lit up over my head, and we returned to the cabin to start our search over. I went inside, found Lola, put her on a leash, and began to walk with her around the grounds, staying close to the bushes in the immediate vicinity of the cabin. Sure enough, within just a few minutes, Roxy crawled out from underneath one of those bushes, perhaps because Lola was there. I'm sure she'd been hiding there the entire time we searched.

> **Lost dog lesson #3:**
>
> Know your dog well, and keep that knowledge in mind when searching. It can help you figure out where your dog might be.

Homing Instincts

The homing instinct of an animal is a combination of their drive to return home and their ability to do so over long distances. It requires some sense of home as well as keen navigation skills. With time, Roxy became more confident and, I believe, came to understand the concept of home. She escaped a few more times but I ultimately always found her waiting on the front step of our house in Boulder, Colorado, panting and looking completely tuckered out. It seemed to me that she had stopped running away to hide, and had instead begun running just to have fun, coming home when she decided her adventure was over.

I'll never know whether Roxy ran in pursuit of a scent or just to explore. While I would panic when she escaped the house or yard, it was always a joy to see her come home on her own. She made enormous progress in bonding to me—or at least to the concept of our house as a home base. If we were out hiking and she ran off, she always found me on the trail. Despite all her running, Roxy never got lost, and I eventually successfully trained her to come back to me when I called her. She would spend hours hanging out underneath a large bush in the fenced back yard, but I think she came to feel that hiding, like running, was something she could do for the sheer pleasure of it, rather than because of fear.

Josey: Visiting Friends

Legends abound about canine navigational abilities. I've observed small miracles myself.

One day while inside the house where I was living in Boulder, I heard Lola and Roxy barking in the back yard. I went outside to get a closer look at what had gotten their attention. They were barking at something outside the six-foot-high privacy fence. I couldn't see through the fence, so I opened the gate that led to the alley. Into the yard ran Josey, a black Lab mix who belongs to my friend Julie.

Just like my friend Melanie had done years earlier when she found Lola, I called Julie to ask her why Josey was at my house. Julie, not at home, had no idea that Josey was gone. It turned out that Julie had gotten a phone call earlier that day from a flower shop that wanted to make a delivery. Since she wasn't going to be at home, Julie told the florist it was okay to open the door to the front porch and place the flowers inside. Josey had never been the kind of dog who would bolt out through an open door. This time, however, she escaped. We don't know whether she was afraid of the stranger who opened her front door, or made a quick decision to visit her canine friends.

Josey, *photo by Julie Teel Simmonds*

Josey had visited my house many times before, and had even stayed with me when Julie traveled out of town, so she felt like a member of the family. But she had never arrived by herself, or through the back gate, or by walking the route between her house and mine. Julie's house was about two miles away, and Julie always parked her car in front of my house, not in the alley. Yet Josey knew the route well enough to navigate by herself, cross several streets, and find her way to the alley and my back gate.

Shanti: Home and Attachment

After Roxy passed away, I adopted a dog I named Shanti. Shanti was a mix of American Staffordshire Terrier, a type of pit bull, and unknown other breeds. He spent seven years in a municipal shelter before I adopted him. I don't know whether it was nature or nurture, but Shanti was the exact opposite of Roxy. He was confident around people and quickly bonded to me. If he wasn't at home, there was only one place he wanted to be, and that was by my side.

Shanti, *photo by Amy Mall*

One summer he joined me on a trip to a small village in Vermont. Late one night, we arrived at the carriage house I had rented. Neither one of us had been there before. The next morning we went for a run to explore a nearby trail. As soon as I reached the farthest point on our route and decided to turn around, Shanti started hightailing it back to the carriage house. He ran ahead of me the entire way until we reached the front door, as if he wanted to get back to the place that seemed like a home base to him. Somehow he knew exactly how to find the way back on his own. I don't think any of my other dogs would have done that.

I had a similar experience with him the first time he hiked with me in Washington's Rock Creek Park. After a few hours of hiking, we were nearing the parking lot where we had started. He pulled me off the trail and crossed a field, stopping right in front of our car. Perhaps I gave him some cues about direction without realizing it, but he seemed to have an excellent ability to navigate, and an uncanny instinct that the car was our destination and our connection to home.

Quint: Instinct or Intelligence?

I met Ann Yoder through her daughter-in-law, a work colleague of mine. Ann told me the story of her dog Quint, a small dog, only 12 inches high, who looked like a Spitz mix. The Yoder family purchased Quint from another family when he was less than a year old. Quint, Ann said, had the spirit of a big dog inside a small dog's body. He was very attached to the three Yoder children, but would also take any opportunity to squeeze out an open door and run around their neighborhood in the Shenandoah Valley of Virginia.

About a year after Quint came to live with the Yoders, the family went on an extended road trip. The Yoders left Quint with Ann's brother and his family, who lived about seven miles away in a small village surrounded by farmland. Quint had never been there before, but it turned out he loved it. Three weeks later, when the Yoders picked him up, they were regaled with stories of Quint's enjoying rural life and chasing rabbits, squirrels, and raccoons. It was clear, said Ann, that he had been "in his element."

Is My Dog Like Lassie?

Quint, *photo by Paul Yoder*

That same day, the Yoders arrived back at their house and began settling in after weeks away. Later that afternoon, they realized that Quint had gotten out and gone missing, just like he had done before their trip. This time, however, Quint didn't come home on his own like he usually did. The Yoders had no idea where Quint was.

The following evening, Ann's sister-in-law called. "Did you know that Quint is out here?" she asked. Quint had showed up at her house within hours of being picked up by the Yoders. He was a foot high, and had traveled that route only once before—by car—yet he successfully navigated the seven miles on his own. Quint clearly loved being out in the country, and the Yoders, as much as they loved him, let him stay. Thinking about Quint's drive and ability to navigate, Ann asked, "Is it instinct or intelligence?"

3

How Wolves Navigate

Wolves have been called the smartest predator in the wildlife kingdom. They are also highly social: wolves form interdependent relationships, work cooperatively, live in groups, and honor their pack bonds.

The pack is a finely tuned structure that has been honed over tens of thousands of years to work highly efficiently to the benefit of every member. Wolf packs generally consist of a male and female pair and their offspring and other relatives. While there is indeed such a thing as the classic "lone wolf," they are usually just temporarily alone—a younger wolf who has struck out on his or her own to form a new pack or join another existing one.

Regardless of whether members of a wolf pack are all genetically related, packs always function as a bonded family unit. Each wolf has a critical role in preserving the pack, whether that role is hunting, caring for pups or ill or injured packmates, or defending against outside threats.

In addition to their functional roles, packmates become playmates, form friendships, and develop other emotional attachments. The bonds between each pack member are essential to keep the pack functioning as it is intended—for the survival of each member.

Wolf Navigation Skills

The gray wolf is the wolf species found in the western United States, the Great Lakes region, and Canada. A gray wolf pack's territory is typically hundreds of square miles, but they've been known to range up to 1,000 square miles or even more. Within their home territory, wolves can roam 50 miles or more a day as they hunt for prey and rotate their hunting sites.

Every wolf pack has a home base within its territory, where its den is located. The den is used in the spring for giving birth and rearing young pups immediately after they're born. Once the pups are old enough to safely leave the den, the pack will start using other sites spread around its territory, known as rendezvous sites, to take advantage of more hunting opportunities. Rendezvous sites are typically located closer to prey migration paths and prime hunting areas. Wolves use them to meet up with their packmates, sleep, eat, socialize, and rest. They use the rendezvous sites in late summer and into the fall. As winter starts and migration ends, wolves become nomadic and wander throughout their territory searching for prey.

While wolves have excellent navigation skills, most never leave their home territory unless they have to, perhaps to avoid a competing wolf pack. Still, while wolves often follow the same trails over and over again, they do more than just memorize pathways learned as pups. Wolves create new trails throughout their territory, and they also invent new shortcuts and detours between their trails. They use these shortcuts to do things like intercept prey, and can still find their way back to where they want to be.

Wolves mark these trails, shortcuts, and detours with scent from their urine or released from glands on their paws. Researcher Roger Peters observed wolves in the wild and found that they weren't randomly marking wherever they happened to wander. Instead, they marked identifiable landmarks, such as a rock at the beginning of a shortcut. They also marked the boundaries of their territory and areas that were special to them—like their rendezvous sites.

In other words, wolves use what are now known as "mental maps" they've formed with their senses of smell, sight, and sound. They use these maps to expertly travel for miles through a wild landscape, always finding their way back to their dens and rendezvous sites. They use landmarks such as rocks, streams, hills, and other characteristics of the local terrain, and have marked them, just like our streets are marked with street signs.

Wolf Homing Instincts

Some wolves have found themselves far from their home range, in unfamiliar landscapes, providing a true test of their navigation skills. In a process known as translocation, wild wolves are relocated to a new area, outside their home range, by government authorities. Translocation is usually the result of human complaints about wolves who are preying on livestock, such as cattle or sheep.

Wildlife managers may translocate a single wolf, a group of wolves, or an entire pack. The wolves are tracked, sedated, and then fitted with radio collars so that their whereabouts can be monitored once they are released in their new location. Their new habitat is selected to have everything they need to succeed, often in a national park or a national forest. It should be far from human settlement and livestock grazing, buffered from any potentially competing wolf pack territory, and abundant with wild prey. Although miles from their original home, in an unfamiliar landscape, the translocated wolves should have no reason to leave.

Between 1989 and 2002, 88 wolves in the western United States were translocated. Some were moved on their own, and some in groups. Some were moved multiple times. In total, there were 42 of what scientists call "translocastion events" that took place in Idaho, Montana, and Wyoming. An analysis of these translocations confirms the incredible homing instincts and navigation superpowers of wolves.

Wolves who were translocated with nearly complete family groups stayed in their new territory. According to Liz Bradley, a wildlife biologist who analyzed the translocation results, "There were only four translocation events that resulted in wolves staying at release sites and establishing a pack. Three of those were cases where almost complete family groups were relocated together. The other case was also a group of wolves that was translocated together, albeit not a complete family group."

Only wolves who were with most of their pack stayed in their new territory. All of the other wolves—those translocated by themselves or in smaller groups—left their new locations, despite being provided the best possible habitat that met all their needs. Many of them successfully made it all the way back to their original home territory.

This is an incredible feat. These wolves were transported from their homes over completely foreign routes—while sedated. They navigated distances from 46 miles to 196 miles to get home through unknown lands consisting of rugged wilderness, lakes, and steep terrain.

There were also wolves who were translocated by themselves or in a small group and who left their new territory but who didn't make it back. These wolves were found somewhere between their new location and their original territory. While they didn't make it all the way back to their original home, it certainly appears as if they were trying to reach the rest of their pack.

4

The Science of Dogs and Navigation

Wolves who have been separated from their packs have climbed mountains, crossed lakes, and navigated blindly over long distances to make it back to their families. Like the classic Motown song says: "....*Ain't no mountain high enough, ain't no valley low enough, ain't no river wide enough, to keep me from getting to you.*"

The behavior of translocated wolves is a testament to the strength of both their pack bonds and their homing instincts. Since the wolf is known as the domestic dog's closest living relative, it makes sense to consider whether dogs share the homing instincts of wolves.

Dogs have extraordinary talents. They've been performing highly skilled jobs, some quite demanding, for hundreds of years. Dogs are trained to use their powerful sense of smell to detect explosives, missing or injured people, criminal suspects, contraband, drugs, cancer, low blood sugar in the breath of someone with diabetes, illegal wildlife parts such as elephant ivory or rhinoceros horn, and more. They guide people with disabilities, alerting them or others to threats, and provide assistance such as opening doors, picking up and bringing items, flushing toilets, turning lights on and off, and dozens of other tasks.

Dogs vs. Wolves: Navigation

While dogs can be trained for specialized assignments, it doesn't mean that they are good at everything, or that they will excel at figuring out how to do things when left to their own devices, like the translocated wolves.

Scientists have designed studies to compare the problem-solving ability of the two species. In one such study, they placed wolf pups raised with humans behind three different types of fences: short (eight feet), long (24 feet), and U-shaped (with sides of 15 feet). There was a window in each fence, and the pups could see food on the other side. To get to the food, the animals simply had to find their way around each fence. When compared to dog puppies who had been subjected to similar experiments, wolf pups were much better at figuring out how to go around each type of fence to get to the food. The dog puppies had many more errors, including sometimes running back and forth along the fence without going around it. When the fence was U-shaped, the dogs made almost ten times as many errors as the wolves.

Other research confirms that wolves exhibit better problem-solving skills than dogs. One study compared a group of wolf pups to a group of Malamute puppies. Both groups were raised with the same wolf foster mom and the same humans. The Malamute breed was selected because of its similarities to the wolf species. The study used four different "puzzle boxes" that had food inside. For each puzzle box, the puppies and the wolf pups had to figure out a way to access the food, such as pressing a plunger in one of the boxes.

The scientists conducting the experiments concluded that "...the wolf pups were both more perseverant and more independent in their efforts," and could often solve a puzzle on the first try. The Malamute puppies, however, often turned away from the puzzles and looked at the human scientists when they discovered that the food was not easily accessible. When they successfully solved a puzzle, it required "extensive experimentation." The scientists cautioned that the results could be due to the age, breed or individual abilities of the puppies. Even so, it is significant that the wolf pups successfully solved the puzzle boxes almost four times as often as the Malamute puppies.

Despite the close genetic ties between wolves and dogs, there are clearly differences in the talents of the two species. Dogs have unique emotional and chemical bonds with humans. They can also be trained for and skillfully execute a range of highly specialized jobs unlike any other animal. On the other hand, the average dog doesn't share the homing instincts, problem-solving abilities, or pack bonds of wolves. Marc Bekoff, who has studied both dogs and wolves, told me that domesticated dogs may have lost some of their navigation ability over time because they didn't need to use it anymore to find food or water.

In their book *The Genius of Dogs,* Brian Hare and Vanessa Woods, scientists at the Duke University Canine Cognition

Center, explain that, while dogs can use landmarks to navigate like wolves, they usually don't. In addition to their own research, Hare and Woods reviewed thousands of scientific papers about dogs and concluded that dogs are "fairly unremarkable" in their ability to solve navigational problems, and shouldn't be trusted to find their own way home.

Lost dog lesson #4:

Don't assume that your dog can find their way home on their own like a wolf can. Some can, but many can't.

The Individuality of Dogs

Although the average dog can't navigate like a wolf, we can't forget that there are individual dogs who have accomplished some pretty spectacular feats, like Quint. Marc Bekoff explained to me, "There are individual differences in a dog's cabinet of capacities." There is no one answer, he said, as to why some dogs find their way home and others don't. Some dogs may be better at creating or reading mental maps.

There's no evidence that all dogs have a homing instinct, or if breed makes a difference, but Bekoff pointed out that there's also no evidence that all wolves can navigate equally well. Just like people, dogs can be talented in one aspect but not in others. Ultimately, Bekoff said, when it comes to understanding why some dogs can navigate home while others can't—or choose not to, "We really just don't know."

Lost dog lesson #5:
Every dog is unique. Don't expect your dog to do what another dog has done.

5

Scout

Despite accounts of missing dogs miraculously finding their way home, there are exponentially more dogs waiting in shelters or reported as lost. If an average of more than a million lost dogs do not find their way home each year, there could be millions of dogs in the United States missing at any one time. It breaks my heart every time I see a "lost dog" sign, wondering where the dog might be and what could have happened.

My curiosity about runaway dogs and their fate turned into something more than casual during the summer of 2009, when I was living in Boulder. On a perfect day with clear skies, I went for a bike ride. Cruising along a bike path, I started noticing sign after sign posted with a picture of a missing dog. Clearly a lot of effort had gone into posting so many signs, and I could tell the dog, named Scout, was beloved and sorely missed. As I rode along the bike path, I felt terrible for Scout and her owner, and steadily scanned the woods on either side—hoping I might catch a glimpse of the missing black dog.

"Little Einstein"

Scout's photo showed a black Labrador retriever and border collie mix with white patches. Her owner, Andrew Newman, was a working college student when Scout went missing. A month or so after that bike ride, I read a newspaper article about Scout's fate. When I began the research for this book, I reached out to Andrew to learn more about Scout's story. He told me that, before she went missing, he'd always thought of Scout as a "little Einstein" because she was extremely smart, easy to train, and loved learning new tricks. Sometimes serious, Scout also had a goofy and playful side. Not only did Scout know how to roll over and play dead when Andrew pulled out a fake gun; but she even learned how to fetch beers out of the refrigerator!

Andrew adopted Scout when she was only eight weeks old. While Andrew had other dogs before Scout, he told me that he and Scout had a different connection, one he described as amazing, a special bond. Scout was reserved around new people, but Andrew said she was attached to him like a "ball and chain."

At the time of her disappearance, Scout was about four years old. She and Andrew had been living with Andrew's father, Robert, for about nine months. They were fortunate to live near open space, including trails, fields, and parks. Even when off-leash, however, Scout always stayed near Andrew and obediently came to him when called.

Robert was spending a lot of time at home due to health issues. When Andrew was at work or at school, Scout and Robert spent many hours together, forming their own special bond. He helped walk and feed her. While their connection wasn't quite as close as the one she shared with Andrew, Scout seemed to sense when Robert was dealing with a flare-up of his symptoms and would go to his side.

Late one night, while Andrew was sleeping, Robert was unable to sleep and decided to take a walk with Scout. The road right behind their home was quiet and led to commercial buildings and open space, an area that was generally deserted at night. Scout was off-leash, and chased a rabbit into the street. At the very same time, a car unexpectedly came around a curve. The car hit Scout, and she took off running away from Robert. Terrified, Robert started calling for her. When she didn't come back right away, he phoned Andrew and had to wake him up with the heartbreaking news that no dog owner wants to hear. It was, said Andrew, "the worst call ever."

The Search

Andrew joined Robert and they searched until daylight without any sign of Scout. They continued their search the next day near the location of the car accident—an area with fields of tall grass where an injured dog might be hiding or might have lain down and died. Walking the landscape in grids, Andrew made sure that every spot was covered.

Andrew didn't find Scout that first day, but that was only the start of his search. He printed hundreds of flyers, like the ones I saw posted along the bike path. He got a pair of waders and loaded up on mosquito repellent so he could walk through the local creeks. For weeks, Andrew spent every waking moment when he wasn't at work searching the areas near their home, along with friends who helped when they could.

He kept telling himself that Scout was smart, had always stayed by his side before the car accident, and knew the area. Andrew had a strong instinct that Scout wouldn't be comfortable leaving the vicinity or straying far from home, and was instead waiting for him to find her, so he kept searching the open space near his house. If the worst had happened, he wanted to at least find her body. It was a summer with a lot of thunderstorms, and Scout hated lightning. Even though Andrew knew that the constant rain would be distressing for Scout and possibly make it harder to find her, he continued his search.

As weeks passed, Andrew would get calls from people who'd seen his flyers. Every call raised his hopes, but none of the leads felt quite right. They were either in a different part of town or much farther away, and Andrew didn't think Scout would travel that far from home. Sometimes the description didn't exactly match Scout. Nevertheless, Andrew followed up on every lead. He visited shelters up to 40 miles away, just in case his instincts were wrong and she had run far. Despite all his efforts, Scout was still nowhere to be found.

Eventually the calls slowed to a trickle. Astonishingly, some people called to ask Andrew to take his fliers down, saying they didn't want to see them on the path anymore while they were running or biking.

One More Sighting

Exactly one month to the day since the car accident, just as he was starting to lose faith that he'd find Scout, Andrew got a call from a runner named Tanya. Tanya thought she'd seen Scout earlier that day along the bike path, in an area that was exactly where Andrew had been imagining Scout might be—within a half mile of home. And this time the description actually sounded like Scout. It was the best tip Andrew had gotten in a month. He was at work when the call came, so he got a coworker to cover for him, and took off to search for Scout once again.

Tanya couldn't meet Andrew, so he had to take his best guess at the location she'd described. A half hour later, Andrew pulled into the parking lot closest to where he thought Tanya had seen the black dog. It had been about five hours since the sighting, and it was a location that Andrew had searched several times before, but he was hopeful that this might be a solid lead.

Andrew spent a little over an hour in the immediate area that Tanya had described, with no luck. He then followed a trail to a nearby place where he and Scout had often walked and played, a spot where a creek and a bike path ran beneath the street. He searched for another hour or so, but still no luck.

Before giving up, Andrew took a moment and racked his brain for any other nearby spot where he used to walk Scout. He remembered one more trail that ran along a different underpass, and decided it was worth investigating this one additional location before he called it quits for the day.

Andrew reached the underpass, scanned the vicinity, and didn't see any evidence of Scout. He noticed three culverts, and stepped closer to check them out. He looked inside the first one and saw nothing. He turned toward the middle culvert to look inside, and there she was. Standing right at the opening of the middle culvert, Scout was staring at him. It was as if she'd been standing there the entire month, just waiting. It was the moment Andrew had dreamed about.

Instinct and Persistence

Scout recognized Andrew immediately, lay down on her side, and then rolled over onto her back. Andrew picked her up and carried her to the car, where she promptly climbed into his lap as he rushed her to the veterinarian. A thorough exam found that Scout had a broken toe and a broken tooth, as well as lacerations on her head, legs, and rear end that might have come from fights with coyotes. She was also malnourished. The vet removed her toe and patched her up. All in all, she was in pretty good shape.

No one will ever know what Scout was doing during that month, but Andrew's theory is that she was so hurt after being hit by the car that she hid in one place until she felt better, periodically leaving to hunt rabbits to survive. He thinks Scout never left the area, and probably heard him calling—particularly when he searched at night, when sound travels best. Andrew believes Scout was afraid to come out until he finally got close enough that she could be absolutely certain it was him.

During his search, Andrew had a strong instinct that Scout would remain close-by because of their powerful bond. Perhaps Scout had an instinct to stay where she did, believing that Andrew would eventually find her. He thinks he ultimately found her because neither one of them ever gave up. They each trusted their instincts, and Scout was found less than a half mile from home.

Happy endings are the best. It's inspiring to know that a month-long search can succeed. But we're still left with a mystery. Scout was hurt, but ran away from Robert, a safe family member, instead of toward him. She stayed in the immediate vicinity, but never responded to Andrew's calls during weeks of searching—even though he is her closest family member and she was attached to him like a ball and chain. And she was near her home, in a familiar location, yet for some reason she couldn't—or wouldn't—find her own way back to the house, where there was food, water, shelter, and her family. Instead she hid for four weeks, perhaps in or near the culvert the entire time.

Andrew's theory is a good one: Scout was scared out of her wits, waiting for him to finally come close enough to know for sure it was him. It's probably the best theory we have. We'll never really know why Scout didn't return home on her own—whether she couldn't navigate her way home, or something made her afraid to leave her hiding place.

Andrew relied on his knowledge of Scout and followed his gut to focus the geographical range of his search and the tactics he used. Using our intimate familiarity with our dogs and trusting our instincts is a crucial element of searching for a missing dog. At the same time, we can't make too many assumptions about our dogs, how they will behave or where they might be, or we risk blinding ourselves to possibilities and limiting our search in a way that reduces our chances of finding our missing dog. There's a fine balance between the two.

> **Lost dog lesson #6:**
> Trust your instincts, and don't give up. Your dog is out there somewhere, maybe close to home.

6

Why Dogs Go Missing

It's hard to fathom why your dog would leave home and not return. It's clear from both scientific evidence and our own experience that there is a powerful emotional bond between our dogs and us, one that often grows stronger over the years. Yet our dogs still go missing from home and family—the very place and people they depend upon for food, water and shelter. There are endless books on how dogs behave when they are with us, but the fact remains that we know next to nothing about what our dogs do when they are on their own, or why.

Anthropologist Elizabeth Marshall Thomas decided to find out. In her book *The Hidden Life of Dogs* she wrote about Misha, a husky. Misha belonged to friends of Thomas, and he stayed with her while his owners were out of the country. Thomas allowed Misha to freely roam the streets of Cambridge, Massachusetts and nearby towns, mostly at night, for almost two years during the 1970s. Her research is controversial, considered fascinating by some and irresponsible by others. Thomas followed Misha, by bicycle, two or three nights each week. Misha loved to explore, and some nights he traveled as far as six miles away. I don't condone the practice of letting a dog roam the streets unleashed, but this is the only evidence I've found regarding what a dog actually does when he's out and about on his own—or thinks he is. Based on her observations, Thomas concluded that Misha wasn't seeking canine companionship, food, or sex, but was motivated to find the spots where other dogs had urinated, and then mark those spots with his own urine.

Why Dogs Run

Dogs are unique in personality, emotions, skills, and preferences. While Misha may have roamed for one reason, there are a host of potential motivations for dogs. Some dogs are content at home but are curious, see an open door or gate, and innocently wander out to see what is going on in the world. They may follow an intriguing or delicious scent, chase another animal, or decide to visit a familiar neighbor. Others may just enjoy the thrill of running. And male dogs, even neutered ones, can be attracted by the scent of female dogs, particularly those in heat.

Researchers at Emory University using MRI scanners found that dogs have "massive" olfactory bulbs—the part of the brain responsible for the sense of smell. They determined that a dog's olfactory bulb takes up about 10 percent of their brain, and remark that a dog's sense of smell is 100,000 times as sensitive as a human's.

Marc Bekoff told me that some dogs might be running away from something they fear or dislike, such as abuse or a stranger in the house. Dogs may get startled and run due to a thunderstorm or other loud noises. Even dogs who have never seemed afraid can turn fearful under certain circumstances. Animal behavior expert Temple Grandin points out that even a dog who is not naturally timid can become frightened by something "novel and unexpected," something that could be a tiny detail. And once fear sets in, it can dramatically change a dog's behavior. According to Grandin, fear can be incapacitating for an animal. "Animals in terrible pain can still function; they can function so well they can act as if nothing in the world is wrong. An animal in a state of panic can't function at all," she writes.

My dog Roxy ran away from a vacuum cleaner and hid under a bush for hours. Scout was terrified by her car accident and ran away from her family into the dark night, staying hidden close-by for an entire month.

> **Lost dog lesson #7:**
> Don't assume your dog is not afraid, even if they haven't seemed fearful before.

Zita Macinanti, director of humane law enforcement for the Humane Rescue Alliance of Washington, D.C., told me that, based on her experience, the most common reason that dogs go missing is because they are in a new home or a new environment. Dogs who are relatively new to their family have not yet developed a strong bond and may not feel attached to their family or their home. In addition, they may not be very familiar with their surroundings, at least not enough to navigate home if lost. Forging a strong

Why Dogs Go Missing

bond can take months or more. The connection continues to strengthen over time, especially with fearful dogs, but it may not be enough to keep your dog from going missing.

The second most frequent reason a dog has gone missing is that they were allowed to be off-leash on a walk or hike, Macinanti said. Even if your dog has been off-leash reliably many times before, they may still take off one day and not come back. Fortunately, Macinanti reports that her team finds 85 percent of the dogs it helps search for.

> **Lost dog lesson #8:**
> Don't walk your dog off-leash unless you are 100 percent confident that you've trained them well enough to come back to you when called.

Why Dogs Don't Return

Just as there are many reasons a dog may go missing, there are as many reasons a dog may not come home once they've gotten out. Marc Bekoff says that some dogs may intentionally run away, but some dogs may just get lost. Zita Macinanti agreed. "Dogs may run from fear or for fun but, either way, some dogs find themselves lost when they finally stop to regroup," she told me.

Anneke Lisberg, an expert in chemical communication signals in dogs, explains that, "While their noses are amazing and they can almost certainly differentiate smells on one block, one tree, even their own footprints, this doesn't mean that they know how to remember those smells in reverse-sequence to trace their way back or know that following the smell of their own feet would lead them back home."

Dogs can also fall victim to a predator, get hit by a car, or become trapped. I can remember two occasions when I found my dog Roxy inside someone else's fenced yard. She found ways to get in, but couldn't get out.

Many dogs will run right up to strangers, especially if there is food involved. While you may think your dog is bonded to you and you alone, or is not particularly gregarious, some dogs will freely go home with other people, particularly if they are hungry, thirsty, tired, or lonely.

It may seem logical that someone who finds a stray dog would try to find the dog's owner or bring the dog to the local shelter, but that doesn't always happen. Even well-meaning people might not take action because they don't know what to do, or they believe the dog doesn't have an owner because it lacks identification. Emily Weiss, vice president of research and development for the ASPCA, told me that, in some cases, people who find stray dogs assume that the dog's owners are irresponsible because the dog was able to get out, and decide to keep the dog because they think they can provide a better home.

Ultimately, there is no one answer regarding why dogs get lost or run away, and why they don't find their way back home. Marc Bekoff told me that, to unravel each mystery, "We need to know the exact circumstances and history of the missing dog."

7

Stolen Dogs

It may seem unbelievable, but dogs get stolen. Reports of pet theft have even been on the rise in some parts of the country. Dogs have been stolen right out of their yards or from a car, and they may be stolen for a variety of reasons. Some thieves may keep a stolen dog as a pet, but others will try to sell a particularly desirable dog, such as a purebred dog. They may use a website like Craigslist to post a stolen dog for sale, and buyers may innocently think they're buying or adopting a dog from its legitimate owner.

Dogs are also stolen to be used for dog fighting. Bernalillo County, New Mexico, experienced a sharp increase in dog thefts from yards and homes in 2015 and, according to the Sheriff's Department, "They're being taken for the sole purpose of being used as bait dogs, and bait dogs are being used in dogfighting rings to allow the champion dog to have something to be able to fight."

There have also been concerns that stolen dogs may be sold to medical research institutions for use in lab experiments. While a 2009 study found little evidence that pets are stolen for research, it also found that it can't be completely ruled out. Fortunately, federal laws have been strengthened since then to reduce the chances that this could ever happen.

In the most disturbing circumstances, dogs have been stolen by animal abusers. The accounts of these incidents are too painful to write about, but I want to provide evidence that they do indeed occur, so I've included one example in the notes for this book.

Maggi

I met Andrew Schneider through a work colleague. Andrew and his wife, Kathleen Best, were living in downtown St. Louis when they brought home a yellow Labrador retriever puppy to join their black Lab, Pepper. They named her Maggi. Their back yard had a six-foot-high security fence and backed up to an alley.

One afternoon, when Maggi was only 11 weeks old, she was outside playing with Andrew and Kathleen's son and two young grandchildren. About ten minutes after the kids finished playing and came into the house, Andrew realized Maggi wasn't with them. He checked the yard; Maggi was gone and the alley gate was open. Upon closer examination, Andrew saw that the gate latch was bent as if it had been kicked in. They didn't know if Maggi had run away through the open gate or if she'd been deliberately stolen.

Maggi, *photo by Kathleen Best*

The Schneider family began combing the neighborhood immediately, calling Maggi for hours. When they couldn't find her, they made posters with her photo and a reward offer, and began placing them on telephone poles and in the windows of local businesses. Starting the next day, they checked local shelters. After a few days with no luck, Andrew found out where the public works department took dead animals and checked to see if Maggi's body was there. Fortunately, it wasn't.

Andrew and Kathleen kept up their search diligently. They periodically got calls from people who said they had Maggi and wanted the reward, but none were legitimate. After holding out hope for about nine months, Andrew and Kathleen brought home an adorable new Lab puppy. They named her Libby, put a huge new lock on their back gate, patched gaps in the fence, and focused on enjoying their new puppy.

A few days after Libby arrived, the phone rang. A man on the other end of the phone said he had just seen a poster in a local pet shop and remembered that he'd seen Maggi months earlier, at a boarding house in the neighborhood. This was the first time that a call about Maggi didn't have a blocked caller ID. The caller said that a young man living at the boarding house had gotten the puppy from his sister, who had gotten her from someone else. The caller wouldn't give Andrew the address of the house, only the name of the street it was on.

Andrew and Kathleen got in their car and started looking for clues. Twelve blocks away, Kathleen spotted what appeared to be a fully grown yellow Lab behind a chain-link fence. The dog came when they called Maggi's name, but they couldn't be sure it was actually her, since they hadn't seen Maggi since she was a little puppy. Thinking creatively, they called Maggi's breeder to ask for help. She put them in touch with the owners of one of Maggie's littermates, and they all met. Seeing the littermate convinced Andrew and Kathleen that the dog behind the chain-link fence was indeed Maggi, and they went straight to the police station.

The desk sergeant asked if there were any identifying marks on Maggi that could prove she was their stolen pup. That seemed like a challenge when the dog in question was a purebred yellow Lab. Andrew thought hard and remembered Maggi's first vet appointment, when the vet had noted that Maggi had two extra nipples. Two police officers accompanied Andrew and Kathleen to the house with the yellow Lab. They called her name, and she came right to them. They all checked her nipples and, finding two extra, the police officers turned the dog over to Andrew and Kathleen, who became the proud owners of three Labs: Maggi, Libby, and Pepper.

Arthello

I contacted Stephanie Felicies after seeing a Facebook post about her dog Arthello. On Christmas Eve 2016, Stephanie arrived home from work. It was late at night and she took Arthello, at the time a five-month-old pit bull puppy, for a walk where they live in the Bronx. Stephanie needed dog food for Arthello and a sandwich for herself. She tied Arthello up outside a local grocery store, went inside to order, and then came back outside to wait with Arthello until her sandwich was ready. A few minutes later, there was a knock on the store window and Stephanie ran inside to pay. There was no line, and Stephanie was quick. When she came back outside, just minutes later, Arthello was gone.

Stephanie told me that she started running up and down the street, screaming out his name. A train had just arrived at a nearby station, and there were a lot of people around. She asked everyone she passed if they had seen someone with a dog. Two women approached her and told her they saw a man dragging a dog who matched Arthello's description. Stephanie ran in the direction they pointed out, but didn't see anyone. She called the police, and a patrol car came quickly; they drove around the neighborhood, but there was no sign of her little pup.

As soon as she got home, Stephanie gathered photos of Arthello and created flyers. She posted notices on Facebook and other websites. Local animal advocates reached out to her and helped her post the flyers starting the very next day. Stephanie spent several weeks walking the streets of her neighborhood, speaking to shop owners, posting more flyers, and reaching out to everyone who crossed her path. There was no sign of Arthello. Then she had a new idea: She would ask local shop owners if she could view their surveillance videos.

Arthello, *photo by Stephanie Felicies*

Incredibly, at one shop, she came across video footage of Arthello and the man who stole him. That seemed like a huge break. When the shop owner refused to give her a copy of the video, Stephanie called a local TV news station.

A TV news crew went to the store and made a copy of the video, then put Arthello on the nightly news.

On her search throughout the neighborhood, Stephanie would meet people who said they had seen a man walking Arthello, and that he was offering the dog for sale. Many had seen the TV news report or her flyers. Every day she walked farther, expanding her search. She staked out one apartment building after someone gave her a tip that he had seen Arthello there, but after a few days it seemed like a dead end. Stephanie got a lot of prank calls and texts—an increasingly upsetting experience for someone who was already very stressed.

On February 12, seven weeks after Arthello was stolen, Stephanie got a call and heard: "Stephanie, I think I found Arthello." She thought it was a joke, because there had been so many prank calls, but Stephanie met the caller, who showed her a picture. It was indeed Arthello! Stephanie bravely got in the caller's car—after taking a photo of the license plate and texting it to her cousin. She wasn't going to lose the best lead she'd gotten in almost two months.

Five minutes later, the woman stopped in front of a house, went inside, and came back out with Arthello. Stephanie learned that the caller's brother had stolen Arthello as a Christmas gift for their mother. The Good Samaritan said she had been looking for information on Arthello's owner and finally tracked down Stephanie by combining information from the TV news report, an illegible rain-damaged flyer, and a lost-pet website.

Fortunately, both Maggie and Arthello were reunited with their families. Persistence, creativity, and courage were essential in finding these stolen dogs. "Don't give up," Stephanie told me, when asked what advice she had for owners of missing dogs. "Stay on point."

Lost dog lesson #9:

Be creative and persistent in your search. Think outside the box.

8

Dora

I was fascinated by the story of a lost dog named Dora after seeing a video of her on YouTube. I reached out to her owner, Kery O'Bryan, to learn more. In 2009, Dora, a stray black and tan German shepherd mix, was found with a litter of puppies in southern California. Dora was thought to be about five years old at the time, and was protective of her puppies and fearful of new people. Because of her fears, she was fostered by a kind woman instead of being kept in the local shelter. With time, Dora became more comfortable around humans, and was adopted. She once escaped her new family's house and turned up at her foster home, a journey that required her to cross several highways—even though she had never traveled that exact route before. Needless to say, Dora was considered a very smart dog with a keen homing instinct!

Dora, *photo by Kery O'Bryan*

Dora was returned by her first adoptive family to the shelter, for reasons unknown. Kery found her there and decided to adopt her and bring her home. Although Dora had separation anxiety, Kery found that Dora learned new things very fast. "I've never known a dog like her," said Kery, who has had many dogs in his life. "She was the smartest dog I'd ever seen."

July 4 Fireworks

In 2010, Kery and his family moved to Sherman, Texas. By then, Dora had become an obedient and playful member of the family. She would periodically escape their yard by jumping over their six-foot-high privacy fence to play in a nearby creek, but she always came home on her own.

On July 1, 2012, the O'Bryan family moved to a new home in Frisco, Texas, about 50 miles away. On the evening of July 4, after only a few days in the new home, Dora was let out into the back yard. Her mom didn't know that the "grand finale" of the local fireworks, with all the loud booms, was about to begin. Likely scared by the explosions of fireworks, Dora jumped over the fence and was gone in a flash. The O'Bryans hoped she'd come home on her own like she always had in their previous home.

The Search for Dora

Dora, however, didn't come home that night. In a panicked effort to find Dora, Kery immediately began putting up signs in the neighborhood and posting notices on Craigslist and local online forums. It wasn't long before he started getting calls from people who reported seeing Dora. Everyone who called, however, said she was being evasive, and no one could catch her, including local animal control officers.

Kery set up traps where she had been spotted, and Good Samaritans who'd read about Dora on local online bulletin boards helped him search. At one point, Kery hired a local pet detective who used search dogs alleged to be trained to follow the scent of a specific individual dog, but they couldn't locate Dora (search dogs are discussed in detail in Chapter 10). As time went on and Dora didn't turn up, Kery kept in mind that she had a microchip, hoping that it would eventually help bring Dora home.

Over time, calls about possible sightings of Dora slowed to a trickle. Then one day, Kery received a call from animal control about a dog that matched Dora's description. The dog had been hit by a car and, when the driver got out of her car to help the injured dog, the dog bit her. Animal control arrived on the scene, captured the dog, and euthanized her before contacting Kery. Kery, however, was certain before even seeing her that the dog in question could not be Dora. "Dora," he told me, "was too smart to get hit by a car." Animal control did not find a microchip in the dog's body, and Kery was convinced that Dora was still alive.

Patience Pays Off

Exactly seven months and one day after Dora ran away from home, Kery got a call from a volunteer at the county animal shelter, letting him know that they'd trapped a stray dog matching Dora's description. An animal control officer had trapped the dog after trying to coax her into his truck for two weeks. This dog had a microchip, and it was indeed Dora herself.

There's a video of Kery going to pick up Dora at the shelter that was posted online and widely viewed. In it, Dora looks bewildered or even stunned upon seeing Kery, and it seems to take her a moment before she realizes that it is truly her dad who has come to rescue her. Fortunately Dora was healthy, even fat, according to Kery, and he believes she survived on rabbits and scavenged trash. She'd been found about 25 miles from home, roughly halfway between her current home in Frisco and the O'Bryan family's former home in Sherman.

Translocated wolves who were relocated with their packs remain in their new homes. So it might make sense to predict that Dora, frightened by loud fireworks, would stay with her family. Instead, Dora ran away, in the direction of her old home in Sherman. While fear may have confused her understanding of where her home and family were, it didn't seem to eliminate Dora's powerful homing instinct and her drive to find them. Kery believes Dora was trying to get to the home she knew best—in Sherman. She'd almost made it.

> **Lost dog lesson #10:**
> Take extra precautions if you have a new dog, you've moved to a new home, or your dog is staying with someone else. These factors increase the risk that your dog will go missing.

9

Top Ten Tips to Keep Your Dog Home

There are steps that every dog owner can take right now to reduce the odds that your dog will go missing. In addition, there are things you can do now that will make it easier to find your dog, and to find them more quickly, should they get lost or run away in the future.

Here are the ten most important things you, your dog walker, or anyone else taking care of your dog can do to keep your beloved pooch safe at home.

1. Walk your dog on a leash, but avoid plain buckle collars: use a harness or martingale collar.

It may seem obvious that dogs should be walked on a leash, yet many missing dog stories involve a dog who was off leash. If you love walking or hiking with your dog off-leash, but your dog is not 100 percent reliable, just don't do it.

It's also essential to have a slip-proof collar or harness for your dog. There are too many reports of lost dogs who ran away because they were able to slip out of their collars while on a walk. This is an easily preventable tragedy. A disproportionate number of these "slipped collar" cases seem to involve newly adopted dogs, foster dogs, dogs being walked by someone other than their owner (such as a friend or professional dog walker), or dogs known to be fearful. If a dog isn't yet bonded to you, they are more likely to run away. Starting the moment you leave the shelter, don't walk your dog with the leash attached to a traditional buckle collar (also known as a flat collar).

Even if these situations don't apply to you, it's still a good idea to avoid walking your dog with a flat collar unless you are 100 percent certain they would never try to get away. It's that simple. Dogs who have never slipped their collars before may still do so in response to a sudden drive to chase something interesting, an unpredictable frightening incident, or another completely unexpected reason.

While you could theoretically tighten a flat buckle collar enough to make it impossible for a dog to escape, it could be uncomfortably tight for your dog. In addition, you'd have to remember to loosen and retighten it before and after every walk.

Sensible alternatives to buckle collars include martingale collars and certain harnesses. Some harnesses are very difficult for a dog to wriggle out of, and I've seen some advertised as "escape proof" or "escape resistant," but others are not as secure. My dog Shanti was able to escape a few harness styles. For example, some front-clip harnesses, designed to stop a dog from pulling, are easy to slip out of. Check any harness closely before using it on a walk. Make sure it is snug and secure. Particularly for new or fearful dogs, consider attaching your leash to a second restraint, using what is known as a "coupler," to ensure your dog can't get away from you on a walk.

Martingale collars are another option for walking a dog. Also known as "limited slip" collars or "limited choke" collars, martingale collars are adjustable and typically do not have a buckle. Instead, they are large enough to slip over your dog's head. The leash is attached to a loop, which cinches a larger loop snug around your dog's neck if the dog pulls. Martingale collars should be customized to ensure that they cinch tight enough to prevent a dog from slipping out of the collar, but not so tight that they could choke your dog. It's recommended that a martingale collar never be left on your dog when they're alone. The collar can be slipped on just for walks.

The use of other dog collars that have a choke function, like a prong collar (also known as a pinch collar) or a choke chain, is very controversial. At a minimum they should never be used unless they've been recommended by a professional trainer, you've fully assessed them for all pros and cons, and you have been thoroughly instructed in their proper use. These collars should never be left on a dog when unattended.

2. Microchip your dog—or consider a tattoo.

Microchips are an invaluable method of identifying the owners of a stray dog. They're especially crucial if your dog has been separated from their collar and identification tag, or never had a tag. And they stay with your dog forever—even if the dog has been stolen or lost for a long time.

Top Ten Tips to Keep Your Dog Home

According to the American Kennel Club, "animals with a microchip brought into shelters have over a 200% higher chance of being returned than non-microchipped pets."

At a typical cost of around $50, a veterinarian will implant a small chip—roughly the size of a grain of rice—beneath the skin of your dog with a quick injection. The microchip is typically implanted between your dog's shoulder blades. If your dog is ever lost and taken to an animal shelter or veterinary office, the staff can use a microchip scanner to detect the chip beneath the skin. The scanner emits radio waves that will activate the chip to transmit an identification number to the scanner. The person with the scanner can then look up the identification number on a registry to find the person who is registered as the dog's owner, along with their contact information.

In one happy ending, the Goldston family was contacted by a shelter nine years after their boxer mix named Boozer had gone missing in Tennessee. Unbeknownst to them, after getting lost, Boozer had been adopted by a different family. At some point he was taken to Colorado, where he somehow ended up in an animal shelter. The shelter scanned Boozer and found the Goldston family's current contact information from his microchip, and they were reunited.

In Florida, Joshua Edwards was contacted by a shelter eight years after his Rottweiler, Duke, disappeared from his backyard. Edwards searched for months, but never found Duke. Fortunately, they were reunited after Duke ended up in the shelter and a scan found his microchip, which was registered with Joshua's contact information.

Microchip technology is clearly effective, but it's not without some complications, so it's helpful to do a little research before selecting the best microchip option for you and your dog. To start, there are many different microchip manufacturers, and there is no one central registry for looking up chip identification numbers; the American Animal Hospital Association (AAHA) "pet microchip lookup" website lists 20 different companies (www.petmicrochiplookup.org). It's therefore possible that someone who has scanned your dog's microchip and tries to look it up will not find your contact information because they've searched the wrong registry. The AAHA website is not a registry itself, but anyone can use it to look up a chip number and find the right registry for that chip. They'll still have to check that registry for the pet owner's contact information.

Second, different microchips operate at different radio frequencies, so not every scanner will work with every chip. There are now universal scanners designed to work with more than one radio frequency, but it's been reported that they may work better with some microchip brands than with others.

Last, there's a small chance that a microchip will migrate from the shoulder-blade area to another part of a dog's body. If a shelter doesn't run the scanner over your dog's entire body when searching for a microchip, it won't find a migrated chip.

Here are some tips you can follow to ensure a successful outcome should your microchipped dog ever become lost:

- Before you get your dog chipped, research which microchip brand will work best with a universal scanner.

- Check with local shelters to see if there is a particular frequency and scanner most commonly used in your area.

- Update your registration every time you move, change a phone number, or add or delete an e-mail address.

- If your pet becomes lost, check your registration immediately to make sure all of the information is up to date.

- Ask your veterinarian to check your dog's microchip at every annual checkup to make sure the chip has not migrated and that it works with a universal scanner. One manufacturer advertises that its chip has a "patented anti-migration feature to help ensure that the microchip will stay in place so that it may be easily located and scanned." That might be worth considering.

Due to complications with microchips, some people may choose to have their dogs tattooed, or to use both a microchip and a tattoo. Dog tattoos are applied on a dog's abdomen or inside leg. These are very visible, are advertised as painless, and also have registries, but are much less common than microchips. I wasn't able to interview anyone whose dog had a tattoo, but more information is available online.

3. Always keep good quality, up-to-date photos of your dog.

If the worst happens, and your dog goes missing, quality photos that are clear and identifiable will be very important to help you find your lost dog. Take photos at least once per year so they are up to date. To start, take close-ups of your dog's face, with different expressions if possible (e.g., mouth opened, mouth closed). If your dog can change their ear positions, take photos of each variation, such as ears standing straight up, flopping down to the side, or pinned back against the head. Additionally, take photos of your dog's full body to capture all colors and markings, the shape of their tail, and general proportions. Don't forget photos of any unique identifying marks.

There are apps where you can upload and store an existing photo of your dog's face, or take a new photo, such as Finding Rover *(www.findingrover.com)* or PiP *(www.petrecognition.com)*. These apps also allow anyone who finds a lost dog to upload a photo, including shelters, veterinarians, or individuals. If you report a missing dog, these apps use the latest in facial recognition technology to analyze your dog's face and see if it is a match for any of the dogs who have been reported found.

4. Train your dog to come when called.

Your dog's propensity to come when called is known in the dog world as the "recall." This may be the most important command your dog will ever learn. A web search will turn up dozens (or perhaps hundreds!) of free articles and videos on how to train the recall command. Another great source of dog training materials is Dogwise *(www.dogwise.com)*. Obedience classes and private trainers are also readily available in most areas for direct guidance. Some dogs are easier to train than others but, regardless of your dog's age or how long you've had them, you can always work on improving the chances they will come when called.

Beyond recall, any positive obedience training can help build confidence in your dog and strengthen the trust and bond between you and your dog. This will promote your dog's feelings of attachment and could help reduce the chances that your dog might run away in the future. Training can be a lifelong pursuit to continually reinforce and enhance a dog's skills, obedience, and relationship with their family.

5. Desensitize your dog to sources of fear.

Some dogs run away due to fear of loud noises like thunder or fireworks, or due to separation anxiety when their family is not home. The puzzle box study mentioned in Chapter 4 was conducted during the month of July, and researchers found that the wolf pups in the study "remained nervous and distressed" as long as 48 hours after July 4, when there were both fireworks and thunder. If you are aware of a specific reason why your dog is at risk of running away, you can use desensitization and counter-conditioning, two tools often used together, to help reduce the risk of your dog's becoming lost.

Desensitization and counter-conditioning are different from obedience training. Obedience training, such as teaching your dog to sit, is intended to elicit a specific behavior in response to a command. Desensitization and counter-conditioning, however, strive to influence a dog's emotional reaction to a situation, such as fear. When combined, they can lead to a new emotional response to something that was previously upsetting.

Desensitization reduces a dog's sensitivity to a stimulus using gradual exposure over time. For example, if your dog is fearful of lawn mowers, you can help them slowly get used to them. The process won't happen in one day. Starting far away from a lawnmower that is turned off, slowly walk closer and closer, seeing how close your dog can get before they start showing any fear. If your dog starts to appear nervous, you've gone too far, too fast. Once your dog is completely comfortable with a quiet lawnmower, repeat the process while someone pushes the lawnmower but it is still turned off. Next, work on getting your dog comfortable with the mower turned on but not moving and, eventually, focus on desensitizing your dog to a moving lawnmower with the engine turned on.

Counter-conditioning goes a step further by teaching your dog to have a positive association with a scary stimulus. For example, if your dog is frightened of the mail carrier, offering a treat or a favorite toy every time the mail carrier comes into view can change their feeling about the mail carrier from negative to positive. Just as with desensitization, it is important to keep your dog under the threshold of reaction, not letting the fear-causing stimulus get too close too quickly. As with obedience training, there are a multitude of resources available to dog owners to get professional assistance with these two effective approaches.

6. Consider a high-tech dog tracking system.

In recent years, inventors have developed high-tech products to help dog owners locate lost dogs, such as tracking collars, or tags that attach to regular collars. The last time I checked, these technologies, which can monitor a dog's location in real time, were available from at least ten manufacturers. Like microchips, however, there are differences among products. Some of these differences are significant. Here are some questions to consider when shopping for a tracking collar or tag-based system:

- Does the system rely on GPS (satellite data), cell-phone coverage, or radio frequencies (similar to walkie-talkies) to track a dog and alert you to their location?

- Does the system allow you to establish a perimeter and get automatic alerts when your dog leaves that perimeter?

- If there is a safety perimeter, what are the minimum and maximum distances from home it can be set?

- If there is a safety perimeter, how quickly after a dog leaves the zone is an alert emitted?

- What is the battery life of the system: days or months?

- How often does the system update?

- Is there a separate collar with a tracking unit, or a tag that clips on to any collar?

- If you have a small dog, does the weight of the unit fit the dog?

- What is the maximum range of the system?

- Once the system is purchased, are there any additional fees, such as subscriptions or a one-time payment for tracking?

A technical tracking system can be an enormous help in finding a lost dog, but doing some research before buying one will help you find the system that best fits you and your dog.

7. Always safely restrain your dog when in the car.

Many reports of lost dogs involve dogs that went missing due to a car accident. Car accidents are frightening. A dog may be thrown from a car, escape if doors fly open or windows are broken, or bolt away from the car when you open the door. Many dogs panic after the shock of a car accident and leave your side. They may hide somewhere nearby or run far from the site.

Safely restraining a dog can help keep your dog inside the car if an accident occurs and prevent them from running away. It can also reduce injuries to a dog. Car restraints are just as important as human seat belts, not only for long drives, but for neighborhood errands. An accident can happen at any time and place.

There are several types of car restraints for dogs, including harnesses that can be attached to the car's seat belt system, crates that are ideally tethered to the car, soft carriers that keep a dog secure, or booster-type seats.

The best solution for you and your dog will vary by your car type, dog size, and other factors. It's difficult to know for sure which system is safest, due to a lack of scientific research. To bridge the knowledge gap, Lindsey Wolko founded the non-profit Center for Pet Safety (CPS) after her dog Maggie was injured in a car accident despite wearing a harness labeled as a car-safety device. When searching for a safer option, Wolko learned there were no regulations or standards for pet safety products, so she purchased several harnesses and hired a crash-test facility to analyze their safety. None passed the test. CPS conducts research and testing, and the organization's website *(www.centerforpetsafety.org)* has valuable information on various methods for restraining your dog in a car. Keep in mind that a harness that serves as a good car restraint may not serve other purposes as well, or may not be the right solution for every dog. I purchased a harness certified by the Center for Pet Safety and ensured it was the right size, but my dog easily wriggled out of it and found his way to the front seat when I left him in the car by himself for a few minutes.

8. Make sure your dog always has visible, legible, and current identification attached to their collar or harness.

It seems obvious: make sure your dog always has some form of securely attached identification (ID). Yet many dogs are found without tags or any other form of ID, or with tags that have outdated information. A 2011 study by the ASPCA found that 80 percent of pet owners believed that it is very important for animals to have ID tags, but only 33 percent of them reported that there were always ID tags on their pets!

Even if your dog has a microchip and you own a tracking device, there is still a critical role for traditional ID tags. Many people who find a lost dog won't take them to a shelter or veterinarian's office to see if there is a microchip, but they might look at a tag and make one phone call to the listed owner. While you might plan to get a microchip at your dog's next veterinary appointment, you can buy or order a dog tag today. Think of it as a "belt and suspenders" approach.

Tags or other forms of ID should have easily legible and up-to-date information, most important your address and phone number. In addition to traditional ID tags, there are many other good options you can purchase, such as small capsules that can hold medical and other vital information on an inserted piece of paper, engraved metal plates that are riveted to collars and lie flat, or fabric collars embroidered with your contact information.

I like to use ID tags that have black print on a white background so they are very easy to read. I've purchased tags that are made of a polymer-coated steel so the identification information won't wear down or get caked with dirt like engraved tags, and that come with a guarantee that they will "never become unreadable" *(www.dogtagart.com)*.

Pick the option that works best for you and your dog, and add a second phone number if you are not always reachable at your primary number. Any form of ID is better than none.

9. When your dog stays somewhere other than home, check the accommodations yourself.

Dogs can escape from a boarding facility, a pet sitter, or a friend's home. They may not be comfortable in a less familiar place, they may just want to go home, or they may think it is fun to run and hide. One way to help prevent this from happening is to closely examine any facility and the surrounding environs before leaving your dog there.

Top Ten Tips to Keep Your Dog Home

You know your dog best: look for possible escape opportunities in the building and escape routes off the property. Check where your dog will be spending its time during the day, as well as where they will sleep.

If you know your dog is particularly fearful or anxious, or if they have joined your family fairly recently and you aren't 100 percent confident about their nature, then it's well worth paying more for the boarding option that is the most secure. Some questions to ask when considering a boarding option:

- Will your dog be left in a crate? My dog Shanti broke out of both wire crates and plastic airline crates.

- What rooms will your dog be in, and are the premises ever left unattended—for example, overnight?

- Are there any windows or doors your dog can access when no one is watching?

- Does the fencing have any gaps, perhaps at the bottom, or can your dog dig beneath it, jump over it, or fit through a crack in the fence?

- What are the protocols when a gate is opened?

- If your dog is particularly small, be extra cautious about gaps in gates and fences.

10. Know your dog.

Whether you're a lifelong dog owner or a first-timer, take the time to know each of your dogs and their individual fears, vulnerabilities, and proclivities. Design your dog's activities and environment around their needs in order to best keep them home and safe. For example:

- If you love walking or hiking with your dog off-leash, but your dog is not 100 percent reliable, just don't do it.

- If your dog is scared of some other dogs or people at the dog park, skip any dog parks that are not completely secure, even if you think it could be fun for your dog to play there. Work on desensitization or counter-conditioning in a safe way.

- If your dog is nervous, or can jump fences or dig beneath them, don't leave your dog unattended in the yard.

- If your dog takes any chance to run out an open door or gate, or has figured out how to open them, develop protocols to ensure that they won't get those opportunities and make sure your fence and gates are secure.

- Don't leave your dog outside alone or walk your dog off-leash on July 4 or days leading up to it, or when there is thunder—even if you don't think your dog is afraid of these noises.

- If your male dog seems to be trying to escape to reach a female dog in the neighborhood, neutering your dog may help (and is the right thing to do to help stop pet overpopulation as well).

> **The dogs most likely to go missing**
>
> Your dog is more likely to go missing if they are:
>
> - newly adopted
> - a foster dog
> - being boarded or walked by someone other than you
> - known to be fearful
> - being walked off-leash
> - left unattended.
>
> Make doubly sure there are no open doors, fence gaps, loose collars, or any other way such a dog can escape from your care.
>
> New or fearful dogs are not only more likely to go missing; they can also be the hardest to find if they escape.
>
> Don't assume that the breed of your dog will determine the risk that they will run away or get lost.

10

Top Ten Tips to Find Your Missing Dog

If your dog has gone missing, you may start to panic. As much as possible, try to keep a level head. Below are the top ten things you can do to find your dog. You may not need or want to do all of them. Some of them will take more time than others. You don't have to do them in the order listed below.

Depending on the circumstances, it is often best to start with a thorough search of your own home and property, to ensure that your dog is really lost. If you can't find your dog close by, then you can move on to posting notices on social media and internet sites. It takes relatively little time to post on the internet, and your missing dog notice can be blasted to many people at once, then quickly shared and forwarded. Using the internet is the fastest way to get people looking for your dog and calling in sightings. But it may not reach everyone, so once you've posted some online notices, you can start to make paper flyers and signs, and then hit the streets to post them and hand them out, making contact with people in your neighborhood. There are more details on these and other tools below.

Finding your dog as quickly as possible will require you to stay calm and take a thoughtful approach. You'll need to consider what you know about your dog, while at the same time not making too many assumptions.

If you are sure your dog is missing, here are the top ten things you can do to help find your dog:

1. Start your search at home and in the immediate vicinity.

If your dog went missing from your house, start by thoroughly scouring your home and all nearby surroundings in case your dog is hiding or hurt. Search under bushes or porches, in sheds or other outbuildings, behind furniture, in closets, under cars, and in every other nearby location.

If your first search around your home doesn't lead to your dog, ask yourself the following questions:

- What are my dog's fears and preferences?
- Where does my dog usually go on walks and what landmarks do they like to visit, such as a local creek or dog park?
- Does my dog like to hide?
- Does my dog ever visit friends or neighbors?

The answers to these questions will help you start the search beyond your property. One study of lost dogs found that the most common way owners found their dogs was by searching their neighborhood. Your dog's personality, appearance, and fears will all contribute to where your dog goes and determining how best to find them, but keep an open mind. Try not to make any assumptions, such as: "My dog would never cross the highway/go to the mean neighbor's house/go home with a stranger," or "My dog can figure out how to get home on their own." Every dog and situation is unique. Jumping to conclusions about your dog in a way that rules out possibilities could lead you in the wrong direction. Act swiftly instead of waiting a day or two to see if your dog shows up on their own. Your dog may be traveling at a quick pace.

If your dog is not quickly found in your neighborhood, they may have traveled farther or been picked up by someone. Landa Coldiron, an award-winning bloodhound handler who works in lost pet detection, told me that friendly and small dogs are particularly likely to be picked up by someone. Coldiron says that missing dogs who are shy or skittish often run away from people and are found in industrial areas, along highways, on golf courses, and in wooded or green spaces. They are often looking for a quiet place as well as a water source. Eventually they will also look for food, although dogs differ in how food-driven they are.

2. Keep track of everything you do.

Before you start, grab a notebook to keep track of your search. You never know when the information you write down will come in handy as you look for your missing dog. It can also be helpful to mark or "pin" this information on a paper or online map so you can visualize your search, including where you have posted signs and talked to people, and locations where your dog has been spotted. Seeing this information marked on a map may help you detect your dog's direction of travel or patterns in their movement. Information you can document in your notebook or on your map includes:

- names and contact information of everyone you meet

- names and contact information of anyone who offers to help in your search

- details regarding all incoming calls

- all sightings of your dog or related evidence, including location, time of day, and any other details

- every time you check a shelter, Craigslist, or other location or resource

- locations where you post signs or ads

- neighborhoods or blocks you have walked to talk to people.

3. Use the internet and social media to quickly begin publicizing, networking, and building a hub for your search.

While many dogs are still found the old-fashioned way with paper signs, the internet is a powerful tool since so many people are now glued to their smartphones, laptops, and tablets. There are several ways to use the internet to get out the word about your lost dog. The first is social networking. Using social networks such as Facebook, Twitter, and Instagram, you can interact with large numbers of people almost instantaneously, and can target people in your area. An online post or tweet can be swiftly shared and shared again, which means more people looking out for your dog, reporting any sightings, and helping you put up signs—even people you have never met.

In addition to social networks, websites such as Craigslist and community blogs serve as modern-day bulletin boards where you can post information about your dog and also look to see if anyone has found a dog with a description that matches yours. Keep in mind that some Craigslist ads will expire, so check every day to see if your post has been deactivated and needs to be renewed.

Here are tips for using the internet to help find your missing dog through social media and networking:

- Gather any photos of your dog. If you don't have any good-quality photos, an excellent idea from the *Lost Dog Recovery Guide* is to search on the internet for a photo that looks like your dog. This could be much more effective than including a poor-quality photo, or no photo at all, on ads, posters, or flyers.

- If you're not accustomed to using the internet, ask friends or family to help you. A smartphone is ideal, since you can get texts or e-mails when you're out searching for your dog. Ask friends or relatives who are social-media savvy to help you post notices and keep in touch with anyone who has spotted or is helping to look for your dog.

- In addition to posting on your own pages and accounts, identify any social media outlets that focus on your community, town, or county, such as blogs, Facebook pages, online bulletin boards, neighborhood networks such as Nextdoor, or community e-mail lists. Ask your neighbors which sites, groups, or lists they use to get information. Join as many of these as possible, and post notices of your lost dog with photos, identifying information, how to contact you, most recent sightings, and how people can help.

- Connect with regional Facebook pages that are dedicated to finding lost dogs, such as Lost Dogs of Texas or Lost Dogs Arizona. Do an exhaustive search to distribute your missing dog's information to as many relevant sites as possible.

- Check ads for found dogs, dogs for sale, or dogs for adoption in the newspaper and online. If a dog has been found, or worse yet, stolen, the finder may advertise the dog on Craigslist or in another classified ad.

- Use social media to help recruit and organize volunteers to help you in your search. Don't be shy about accepting assistance from well-meaning individuals who want to help. The more people out there who can help search, post signs, and network, the greater the chances of finding your dog.

- If you already have your own Facebook page, consider creating a new Facebook page just for people who are helping in the search for your dog to keep in contact with each other, and to organize sign-posting events and other search-related activities. Keep all information updated as much as possible to let people know you're still searching and keep them engaged. This way, your contacts can send the information to their contacts, and your network can expand exponentially. You can also pay to have your Facebook post "boosted" to Facebook members in a targeted geographical area.

- Post notices on free internet sites that list lost dogs (there's a list in the Appendix). If there were only one such site, everyone would go to the same place and you wouldn't have to check many different sites. But there are many of these sites on the internet, and unfortunately there is no way to know which ones will be successful in helping reunite you with your dog. It can't hurt to post your dog on these; one of them helped reunite Stephanie Felicies with her stolen dog Arthello.

- While many people now get their information online, traditional newspapers and radio are still important outlets, particularly in rural areas. Contact them to see if you can post an announcement or ad.

Social media really works. I contacted Amber Yaw after reading an article about her dog Lucy, a beautiful red-hued golden retriever who Amber adopted in 2010 after returning from deployment with the Army National Guard in Kosovo. In 2011, Amber and her parents were living in rural Alden, Minnesota, a mile from the nearest neighbor, and their three dogs were often left outdoors. Lucy was "the sweetest thing in the world," Amber told me. "I've never had a dog like her."

Amber says Lucy hadn't ever tried to run away, but one day when Lucy was two years old she went missing from the yard. Unfortunately, Lucy's tags had outdated contact information for Lucy's previous owner. Amber posted on her Facebook page that Lucy was missing, and within two days someone posted that they'd seen a man at the local gas station with a dog who matched Lucy's description. Amber went to the gas station and asked to review their security video footage. They did not have any outdoor video, but Amber reviewed their indoor video. She found footage that showed a man speaking with the gas station attendant inside the station and showing him a dog collar, but there was no visible identification for that man or his vehicle.

After five days of searching for Lucy with no luck, Amber set up a new Facebook page dedicated to her search for Lucy, called "Help Lucy find her way home." She posted on the Facebook page of every animal shelter, rescue group, and lost-pet group she could find in Minnesota and two neighboring states, asking people to "like" and "share" the page. In less than two hours, someone posted on Lucy's Facebook page that they'd seen an ad on the Minneapolis Craigslist describing a found dog who sounded like it might be Lucy. Minneapolis was about an hour and a half away. Amber used Google to find the Craigslist ad for a found golden retriever, and contacted the person who had posted it.

It turned out that the man in the gas station video had picked up Lucy in the median of the interstate, about 1.5 miles from Amber's home. He didn't know Lucy was local to the area, so he brought her with him to a hotel in Rochester, Minnesota, where he was spending the night. At the hotel, he just so happened to run into a woman from Indiana who had a golden retriever with her. He explained how he'd found Lucy and wanted to reunite her with her owner. The woman at the hotel offered to take Lucy home with her and her dog to Indiana, promising him that she would find Lucy's owner. She was the one who had posted the ad on Craigslist.

When Amber called the number in the ad, Lucy was with this kind woman in Indiana, three states away. Amber believes that social media is an important way to find a lost dog because it can reach so many people and, unlike a sign that someone drives by quickly, someone can see your phone number and your dog's photo over and over again. "I don't think we would've found Lucy without the internet," said Amber.

4. Post lots and lots of visible signs as soon as possible.

Once a dog is known to be missing, it's essential to start posting signs and posters as soon as possible. People who are out and about, those who are most likely to see your dog, may see a sign before they see a social media post. Reports of sightings can help determine where a missing dog might be, the direction they're traveling, and their speed.

Top Ten Tips to Find Your Missing Dog

There's a host of collective wisdom to be found on the internet regarding how to design the most effective signs. If you're not particularly adept at creating a poster, there are free templates online, or services that will do it for you. Resources listed in the Appendix offer guidance on making your own as well as templates that make it easy for anyone to create a basic sign with the most effective characteristics.

Strategies for signs that get the most attention:

- Post larger signs made of neon-colored poster board at high-visibility locations, such as intersections where there is a stop sign or a traffic light.

- Post smaller signs, such as on 8.5 x 11 paper, around neighborhoods and in shops. For outdoor placement, use plastic sleeves with the open side down to protect the signs from the rain.

- Write in large, black, block letters.

- Include a good close-up photo of your dog. As mentioned above, if you don't have any good-quality photos, an excellent alternative is to search for a photo that looks like your dog on the internet.

- Include any key pieces of information, such as a major identifying characteristic, fears, and a familiar command that might help, but make sure the sign is not too crowded. You can make separate flyers for handing out to people and those can contain more details.

- Use staple guns for attaching signs to wooden telephone poles, and "all-weather" clear plastic tape for metal poles, to best withstand the elements.

- Post signs in local businesses and on bulletin boards.

- Consider omitting a key piece of information about your dog so you can determine if a caller really has your dog or is a scam artist.

- Consider whether to offer a reward.

It's important to post a lot of signs, then post more, then keep posting. Do not be shy about accepting offers of assistance and even recruiting volunteers to help you with this task. You need to get the word out as far and as fast as possible. The more people helping, the more you can accomplish. Start posting signs where your dog was last seen, and then work your way out in a radius. Once you start getting reports of sightings, you can focus sign-posting in the areas where credible sightings have been reported. If you're not getting any calls about sightings, continue to post more and more signs farther and farther out.

Kat Albrecht, the founder of Missing Pet Partnership, recommends "tagging" your car windows with fluorescent window markers, a very creative idea. A link showing how to do this is in the Appendix. *The Lost Dog Recovery Guide* also recommends banners, commercial-type signs, and business cards.

The book *Huck* tells the story of a lost toy poodle and his family's desperate search to find him. Huck's owners were on vacation and he was staying with relatives when he escaped through a gap in their fence. He was new to the relatives and their neighborhood. Perhaps for that reason, he ran away when he had a chance.

Huck's family made hundreds of color flyers, posted them in shops and schools, and handed them out as they walked around nearby neighborhoods. They got assistance from many friendly people they met during their search who offered to distribute the flyers more widely and help in other ways. Over a three-day period, they received calls from people who'd spotted Huck and seen one of their signs, which helped them focus their search. Ultimately, someone who'd seen one of their signs nailed to a telephone pole called them with the tip that finally helped them find Huck safe and unharmed—72 hours after he escaped.

Once you have successfully found your dog, please try to remove all the signs you posted.

Should you offer a reward?

It may not always be helpful to offer a financial reward for your missing dog. Landa Coldiron, a lost pet detection expert, recommends rewards only for some dogs, such as friendly dogs who are likely to willingly go to a stranger, or smaller dogs who are easier to catch. Friendly or small dogs are often the kinds of dogs that people like to keep, so they may only give them up when there is a financial reward. A purebred dog that could be sold might also be tempting to keep, and might warrant a reward.

On the other hand, Coldiron told me that she does not recommend offering financial rewards for shy or skittish dogs, because the worst thing would be for someone to chase such a dog in the hopes of capturing them for a reward. This would just make a skittish dog more fearful and cause them to run farther or hide. In these cases, she recommends writing "Do not chase" on the poster, and asking instead for calls or texts with the dog's location. Another option with a fearful dog is to offer a reward "for information only" rather than for the dog itself (while still noting "Do not chase").

If you do offer a reward, be prepared for phone calls from people looking for money who do not actually have your dog. Every lost dog owner I spoke to said they received scam calls. Ask specific questions to determine if someone really has your dog or has seen your dog. If your dog is found, do not give anyone the reward until the dog is actually in your hands. If you meet someone with an intention to give them money in person, it's best to do so in a public place, preferably in daylight, and bring a companion.

5. Check all shelters in the vicinity, and keep checking.

A stray dog may be brought to a shelter even if they have identification tags. It's essential to keep checking local shelters for a lost dog. Here are some important things to keep in mind when searching for a missing dog at a shelter:

- Always go in person. You can call the shelter and describe your dog, but the person answering the phone may not know every single dog, may not think "medium-sized" means the same thing you do, or may not have the same idea of what your dog's breed looks like. It's very important to look in person.

- Ask to see every single dog at the shelter, including any dogs held in back rooms, observation or quarantine areas, medical rooms, outside, or anywhere else. Don't be shy about asking to check every part of the shelter.

- Check the shelter's website, where they might post dogs being cared for in foster homes or who are not at the shelter for some other reason.

- Leave flyers at the shelter, and talk to workers there about your dog. Some shelters also have a lost-and-found dog database where you can post your dog and they post found dogs, but do not rely on these methods alone; continue to check the shelter to view all dogs in person.

- Keep checking back; it may take weeks for a dog to show up at a shelter, perhaps because someone is keeping them at their house or because they have been able to evade people and capture.

- Check with local rescue groups as well as shelters. Some rescue volunteers visit shelters often to look for dogs they can rescue, so they may be able to help look out for your dog. They may also take in stray dogs that never make it to the shelter.

- Check with local veterinarians as well as shelters. Sometimes people will take lost dogs to a veterinarian's office instead.

- Extend your search to shelters outside of your own area. If your dog has traveled for miles, or if someone picked them up and transported them, they may be in a shelter in another town.

- Check back regularly. Some shelters do not hold dogs for very long before putting them up for adoption or euthanizing them. New York City municipal shelters, for example, hold dogs for only 72 hours before deciding whether to euthanize them or place them for adoption. After that, they could still be euthanized or adopted by someone else at any time due to limited space or another reason. Some dogs stay in the shelter longer than that, but there is no guarantee.

Since many dogs evade capture for weeks or months, continue to check until your dog is found. Keep in mind that your dog may look different from when they went missing. Augustus, a stray dog that was rescued by Heart of Alabama: Save, Rescue, Adopt, had a case of severe mange.

It's obvious from seeing his before and after photos that his owners may never have recognized him if they'd seen him in the shelter untreated.

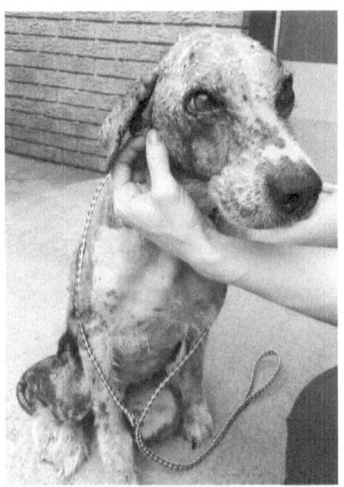 

Augustus, *photos by Heart of Alabama: Save, Rescue, Adopt; hasra.org*

As difficult as it may be, find out how your local municipality handles dogs that are found deceased, such as those that were hit by a car. There may be a log or other records that will help you identify if your dog is, unfortunately, one of them.

If you don't find your dog in a shelter, don't lose hope. Many dogs are found somewhere other than a shelter.

6. Reach out door to door, person to person.

Posting signs and using the internet are essential, but may not be enough. Your search for a missing dog needs to include direct contact with as many people as possible in your target area.

As Kat Albrecht, a former police officer and search dog handler, told me: "You are looking for a person, not for the dog." In other words: You are looking for a person who has seen your dog and can tell you where your dog is. Think about your search as an effort to reach as many people as possible in order to find a person who has spotted your dog.

- Start by creating flyers. These may look like your signs, but can also have more detailed information on them and can use both the front and back. Make sure every person you meet gets a flyer with your contact information.

- If you're in an area where a non-English language is frequently spoken, you can print the information in the other language on the back side of the flyer.

- Make many copies.

- As you walk the neighborhood where you think your dog is missing or where they were last seen, knock on every door and explain to people that you are searching for a missing dog. It may be best to do this during morning or evening hours when people are more likely to be home and fellow dog owners are out walking their dogs. You can use the middle of the day for other search activities. If people haven't seen your dog, ask them to keep an eye out and to contact you if they spot your dog or encounter any clues.

- Make a list of everyone who may regularly visit the neighborhood where your dog is missing, such as mail carriers, UPS and FedEx truck drivers, trash services, school-bus drivers, police officers, and landscaping companies. Contact them to ask for their help. Provide them with a stack of flyers to share with colleagues who might also be driving around your neighborhood.

- Look for residents who regularly run or walk through the neighborhood, including Neighborhood Watch groups, individual runners and bikers, dog owners, and dog walkers.

- Impress upon everyone you contact how important it is to you and your family to find your dog.

Rain's story

My friends Randi Spivak and Andy Kerr adopted their shepherd mix, Rain, when he was about five months old. Rain was always friendly and playful, obedient with commands, and well socialized with people and other dogs, spending his days at doggie day care. One thing Rain has always hated, however, is being in a crate.

In the summer of 2015, Randi and Andy were visiting Boise, Idaho. Rain was about five years old and was going to be boarded for one night in a local doggie day care facility. The next morning at 7, Andy got a call from an employee at the facility. He'd arrived that morning to find Rain's crate open, papers strewn around a desk in the office, a window screen pushed out of its frame, and Rain missing. It looked like Rain had busted out of the crate, jumped up on the desk, escaped out the window, and jumped over the back yard fence.

Rain, *photo by Randi Spivak*

Rain's owners hadn't known that he would be kept in a small crate overnight, or that the facility was left unattended. He'd jumped out the window into a yard with a six-foot security fence, but there were some planters alongside the fence that made it easier for him to escape.

Another employee had already started searching the neighborhood and saw Rain at a gas station down the street. Randi and Andy raced to the gas station, but by then he was gone. They then began a more concerted search.

Top Ten Tips to Find Your Missing Dog

They printed hundreds of flyers and began posting them on telephone poles in the immediate neighborhood. They also handed them out to anyone they encountered as they walked and drove around. Several people offered to help and became part of their core search team. A clerk at a local convenience store took a stack of flyers to hand out to customers, and a traffic-control flagger at a construction site began handing them out to truck drivers.

That first evening, Andy contacted a company that makes recorded phone calls to local homes, and arranged for phone calls to 5,000 homes. The next morning, phone calls started coming in. Some were sightings, some were offers to help, and others were people saying they were praying for Rain. As Randi and Andy walked neighborhoods again the next day, many people they approached mentioned that they'd gotten a phone call about Rain.

They pursued every reported sighting that day. Twice they went to see dogs who people had found, but they weren't Rain. At one location Randi spotted Rain, but he wouldn't come; instead he ran so fast that she lost track of him. Other times Rain was gone by the time they reached the location.

After two days of searching, Randi and Andy upped their game. About ten local residents had volunteered to help, so they bought a map, made copies, and marked territories for each one of their volunteers to cover by posting signs and handing out flyers. Every time they got a new reported sighting, they tackled that neighborhood with posters and flyers. In one neighborhood, Randi and Andy gave flyers to a mail carrier and a UPS driver in their trucks. They also continued to check the shelter.

On Day 4, calls were starting to drop off, and energy was flagging. Randi and Andy contacted a local search-dog handler who could meet them that evening at the location where Rain was most recently seen. The search dog followed a scent past a freeway on-ramp, to a different neighborhood, and then to a local hotel, where he stopped and indicated there was no further evidence of Rain's scent. Another UPS driver was at the hotel, so they told him about Rain and gave him a flyer.

On Day 5, Andy and Randi continued posting flyers in additional neighborhoods near the hotel. At around 5 p.m. they got a call from a UPS driver who'd heard about Rain from the UPS driver they'd met at the hotel the previous day. "I see your dog right now," he said. Randi and Andy rushed to his location, a light industrial area. They found a forklift driver who had recently seen Rain, but Rain was no longer there.

Then another call came in. It was from a woman who said her daughter had just seen Rain, but it was a long way from their current location and Andy thought it might be another wrong-dog sighting. Although doubtful, Andy and Randi were about to speed off to check out this latest lead when another call came in.

This time the call was from two postal workers who were in a car on the freeway. They had seen Rain's flyer near the time clock at work, where it had been posted by the mail carrier Randi had met two days earlier. "I see your dog right now, he's on the westbound side of the freeway, running east," one of the men told Randi. Fortunately, they had remembered the flyer when they saw Rain on the highway, and had called their office to get the phone number on the flyer. The two men, who were driving eastbound, got off the freeway and then got back on going in the other direction. Randi asked them not to call Rain, since he was panicked, but to keep him in their sight. Andy drove there as fast as possible through slow-moving rush-hour traffic.

As they approached the on-ramp, Randi and Andy saw one of the postal workers who'd gotten out of his car to follow Rain. Then they saw Rain himself, trotting directly toward them. Terrified Rain would run away again, Andy got out of the car, opened the door, and calmly said, "Rain, get in the car," as he might on any other day. Rain leapt into the car, as if everything were normal, and Andy quickly closed the door. The power of flyers brought five days of non-stop searching and panic to a happy ending.

7. Consider paid services that publicize your missing dog.

There are companies that offer a host of fee-based services to publicize your lost dog with neighbors and businesses in your community. Their services may include: creating and posting signs or other types of awareness materials; placing automated phone calls to phone numbers associated with addresses within a certain distance of your home or wherever your dog was last seen; posting and monitoring targeted Facebook ads, Craigslist notices, or Twitter tweets; faxing notices to shelters, rescues, and veterinary offices; sending postcards to nearby addresses; or responding to sightings. In addition, the U.S. Postal Service's Every Door Direct Mail Service will mail postcards or flyers by zip code.

A list of some of these companies is included in the Appendix. These services offer a way for you to target people who live in the area where your dog was last seen but may not see a sign or social media posting. They can be especially helpful if you can't corral a group of volunteers, or are searching in a rural area. Some of these services post success stories on their websites, but I wasn't able to verify them. Online reviews are mixed. Randi and Andy used one of these services in their search for Rain, and it was clear that many people had gotten the phone calls that they paid for. On the other hand, a news investigation into one company found that it didn't contact all local businesses as promised and charged a dog owner for calls to non-working numbers. So buyer beware: Consult with the Better Business Bureau and other online rating sites.

8. Take additional steps if your dog has been spotted but won't come.

Once you have a credible sighting of your dog, if you're lucky, you and your dog will rush to each other for a happy reunion. The reality, however, is often a much different scenario. If your dog was shy and skittish to begin with, or if they have become fearful while running or lost, your dog may be in a panic and may not come when called. Or they may think the chase is part of a game. If your missing dog has been spotted but won't come to you, here are steps you can take:

- Stay quiet, and try not to run up to your dog or chase them—this may cause your dog to run away from you. Get low, such as crouching, sitting down, or even lying down, and call gently to your dog. Be patient; it may take a long time for your dog to come close to you. Try not to startle your dog, and remain calm.

- If your dog is bonded with another pet at home or the dog of a friend or neighbor, bring the other animal with you on your search. Let the other pet be the one to approach the missing dog. This worked for me when Roxy was hiding. It might help to have the leashed dog or cat on a long lead (20 or 30 feet).

- If your dog is the playful type, they may think that "chase" is a game, like tag, and may follow you if you run away from them. Make sure your dog is watching you before you start to run, and don't lose sight of your dog, in case the game doesn't help you capture your dog. Dog trainer Adrienne Hovey told me about another game for dogs who won't come that she calls "the most fascinating blade of grass in the world." She suggests you crouch down wherever you are and pretend to examine an imaginary spot on the ground like it's the most amazing thing you've ever seen and, out loud, exclaim something like: "Oh, wow, look at this! I can't believe it!" Hovey says that many dogs will come running to see what you've found, and allow you to grab them.

- Some dogs just love car rides or feel like the car is a safe place. Leave a parked car with an open door to see if your dog will hop in.

- Try using food and/or a worn clothing item with your scent on it to lure the dog to you or your house. If you think your dog is hiding somewhere near your house, you can put a dog bed, a piece of your clothing, or food and water outside your house. Food may attract other animals, including wildlife, but it's worth a try if nothing else has worked yet.

- If all else fails to get your dog to come to you, you can use a humane trap. These are often used for dogs who are skittish or evading all other methods for some other reason. Local rescue groups, animal control agencies, and shelters often have traps they can lend you. If not, you can purchase one. Ask for assistance from trained personnel or volunteers who can help you set it up properly to capture your dog.

    Many factors contribute to successfully trapping a pet, including the size and type of trap, the location you select, the type and placement of bait inside the trap, and how the trap is set. Do not be shy about asking for help; there are people out there with experience who want to help you. Some of these groups will also have cameras so you can monitor the site to see if your dog has approached the trap. Traps need to be checked regularly to release any captured wildlife or rescue any captured pets, replace bait, and check images if there is a camera. More details on trapping methods are available on the internet.

Top Ten Tips to Find Your Missing Dog

- If you've spotted a stray dog who belongs to someone else and won't come to you, it's best not to chase the dog, but to try to find a flyer or social media posting with a number you can call to ensure the owner is on their way to the site. Then remain quiet and monitor the dog's location until the owner arrives. If you can't find any owner information, you can call the local animal control agency and post on various social media or other internet sites that you've found a dog.

After a 2014 car accident in Stowe, Vermont, Kirstin Campbell opened the door to let her golden retriever Murphy out of the car. Unfortunately, Murphy took off running. His family went door to door to hand out flyers in the area, set out food and worn clothing, posted notices on internet forums, and even consulted with a pet psychic. Thanks to all these efforts and help from local volunteers, Murphy was spotted from time to time up to 15 miles away from the accident site, but he would never come when called.

More than a year later, a man spotted Murphy behind his house in Waterbury Center, Vermont, and began feeding him. Kirstin's grandfather built a trap that seemed foolproof, and one night Murphy entered the trap and was captured, but he managed to quickly chew his way out of it. A normally friendly dog, Murphy was still in panic mode. After that escape, the trap was fitted with a different trigger mechanism and finally, in January 2016—more than a year and a half after the car accident—Murphy was captured for good. No one knows how he was able to survive a bitter cold Vermont winter that saw record temperatures as low as -30° Fahrenheit, but fortunately a humane trap finally allowed Murphy to be reunited with his family.

9. Consider hiring a pet detective.

Most people had never heard of a "pet detective" until 1994, when the film *Ace Ventura: Pet Detective* hit the silver screen. It turns out that pet detectives really exist, and they really can help locate missing animals. Services that might be provided include: creating and putting up posters; handling social media outreach; phoning and faxing local businesses and animal hospitals; taking incoming calls and weeding out potential scams; responding in person to sightings; canvassing neighborhoods with flyers; setting up traps and camera stations; organizing groups of volunteers; maintaining maps of sightings; and more. Of course, these services are provided for a fee.

You may not have the time or ability to do these things yourself. Even if you do, the skills of an experienced pet detective should increase the effectiveness of your search. They should be able to discern clues and evidence, and be familiar with how best to deal with skittish dogs. They may also have specialized equipment, like traps or motion-sensor cameras. If there is no pet detective near you, some of these services can be provided remotely, such as consultation services, social media outreach, or advice that can increase the likelihood of a successful search. Pet detectives may also have dogs who are trained to search for other dogs.

Anyone can call themselves a pet detective, since it's not a licensed profession, and each pet detective might offer different services, so research a detective's experience and skills before deciding to hire one. The Missing Pet Partnership *(www.missingpetpartnership.org)* has a lot of helpful information on its website, including a directory of individuals who have completed training in missing pet detection, but they do not endorse or certify the competence of any of the pet detectives listed on their website "beyond completion of the training."

In addition to pet detectives, there are also people who advertise themselves as animal communicators or pet psychics who can help locate missing dogs. I was not able to find anyone to interview who had found such services to be helpful in a missing dog search.

10. Consider hiring a search dog team.

Dogs have been trained to search for humans for more than a century, making canine search and rescue (SAR) a well-established profession. The scent detection skills of a well-trained SAR dog are incredible. Chris Boyer of the National Association For Search And Rescue, which provides education and credentialing for human search and rescue, told me that SAR dogs detect scents that come from skin oils and skin particles that are blown in the air and land on the ground, grass, pavement, or bushes.

Dogs detect the scents of specific individuals in two ways. They may be "tracking" dogs, which means they follow scents on the ground made by footprints within hours of someone going missing, hopefully leading directly to the missing person. Or they can be "trailing" dogs, which means they detect scents in the air or on the ground—up to a few weeks old—to indicate the general direction a person has traveled. This may not identify the exact path that a person has traveled, but it will narrow the search.

The detectability of a scent can depend on many factors, including the age of the scent, intensity, temperature, level of moisture in the air, and where the scent landed, such as the terrain and the type of surface. For example, a scent on wet grass will be detectable for a longer period of time than a scent on dry pavement.

It's important to note, however, that "SAR" is a term reserved only for dogs who search for humans and human evidence. There are reputable training and certification programs for SAR dogs and handlers, as well as regulations that vary by state.

In a relatively new twist, dogs are now being trained to search for missing pets. There isn't one name universally used for these types of dogs, so they may be called "pet-finder" dogs, "missing animal rescue" (or "MAR") dogs, "pet detection" dogs, or another term. I am going to call them pet-finder dogs.

Pet-finder dogs reportedly have helped locate many missing pets. Bloodhound handler Landa Coldiron told me that, even if a pet-finder dog does not find the missing dog, they can lead to valuable evidence, other information regarding the missing dog's direction of travel, and eyewitnesses. These clues all help target a search area. On her website, Coldiron has posted hundreds of fascinating maps that show the routes her search dogs have taken while following the trails of missing pets *(www.lostpetdetection.com/Scent-Trail-Maps.html)*.

Pet-finder dogs are more likely to be helpful in a search for an elderly or ill dog who can't travel far, a search where there have not been many sightings reported, or a situation where the missing dog is fearful and may be hiding. According to pet detective Danielle Robertson, "Very few lost dogs are actually found and captured during the search" by a pet-finder dog. Search-dog handlers should provide an honest assessment about how helpful they can be in your particular circumstances.

There aren't currently any standardized trainings, certifications, or regulations for dogs that search for other dogs. Bill Dotson of the Search Dog Organization of North America, which works with SAR dogs only, told me that a qualified pet-finder dog would have to be trained specifically to search for a dog's scent, including how to distinguish a lost dog's scent from any other dog scents that may be in the area, and their handler must also have appropriate training. Mary Helinski of the North American Search Dog Network, another SAR organization, told me that the key to the success of a search team is high quality, targeted training for both the dog and their handler. "Quality is always more important than quantity when training," she said, "although both are needed."

Since there is no standard training or certification of pet-finder dogs to indicate whether a dog handler and their dog are properly trained, and there is typically a fee for their services, it's important to exercise due diligence before hiring one. Some owners searching for missing dogs have reported spending a lot of money with no results and a lack of confidence in the person they hired. Not all dogs are going to become excellent search dogs. Here are some steps to determine whether a pet-finder dog and their handler are likely to be able to help in your search for your missing dog:

- Speak with the dog handler who will actually be conducting the search. Don't make arrangements through a third party or by e-mail.

- Ask the dog handler about their experience, training, and qualifications, as well as their dog's, and how they continue to train with their dog.

- Ask for at least three recent references and talk to at least two. Contact these references and ask them exactly what the team did and how they carried out their work.

- Ask if they ever turn down cases and, if so, why. Do they ask you about your dog and the circumstances of their disappearance, or does it seem like they take any case no matter the circumstances? A handler who seems to take any case may not be realistic about their team's abilities.

- Ask whether they will be able to tell you what happened to your dog even if they don't find them. Search dog handlers should only communicate the facts to you, and not theorize about the fate of your dog. They also shouldn't offer any guarantee.

- Sign a written contract with specified services and payment requirements.

11. Don't lose hope.

Okay, I have more than ten tips. But this one is really important. Dogs have been found days, weeks, months, and even years after going missing. Kat Albrecht writes: "Physically, your dog is *somewhere* and it did not vanish from earth!"

You've already taken critical steps to increase your chances of finding your dog by reading this book and determining the most effective ways to bring your dog home. If you don't have the money to hire a pet detective or other private service, there are still many things you can do on your own. Don't hesitate to ask for volunteers or accept help from others. Dog lovers are a passionate and giving group—there are many kindred spirits out there who will volunteer to step up and assist you in your search.

Continue to search for your dog. Explore additional ways to publicize your search and reach people who may have seen your dog. Persistence and creativity will serve you well. Don't ever lose hope that you'll bring your dog home.

Epilogue

Grieving a Missing Dog

While I urge you to not give up hope on finding your missing dog, there may come a time when you can't help but grieve, even as you still search. I've never had the experience of a lost dog who never comes home. I can't imagine the sorrow.

In her blog *Grief Healing,* Marty Tousely writes that grieving a missing pet can be harder in some ways than grieving one who has died, because of the uncertainty involved. You don't know whether your dog is alive or dead, suffering or being cared for by someone else. Tousely describes it as a "wound that cannot heal." This grief can be accompanied by deep feelings of guilt or shame if you feel that you are somehow responsible for your dog's disappearance, or that you could have taken steps to prevent this tragedy.

I met Jon Luke when I was writing this book. He told me the story of his Newfoundland Ursula, a sweet, beloved family member, nicknamed Ursy-bear, who always stayed

Epilogue — Grieving a Missing Dog

close to home and never wandered. Jon had brought Ursula home when she was a puppy, and she had been in his wedding. She had lived a long life for a Newfoundland, to 13 years old, when she disappeared one day from the family's securely fenced back yard on the coast of Oregon.

Jon and his wife were racked with panic. They searched intensely, worried about her being out in bad weather, and feared that she might have been attacked by a mountain lion, or been stolen, or encountered some other terrible fate. They searched for a very long time, hired an animal communicator, posted flyers, drove around different neighborhoods, and talked to everyone they encountered. The only calls that came in were false leads. Ultimately, they never found out what happened to Ursula.

Ursula went missing more than ten years ago, and it's still tormenting and tragic to think they will never know what happened to her.

Please take steps toward healing your grief. You can't take too many. Have compassion for yourself, and practice self-care. Support is available from others; don't hesitate to ask for it. In addition to individual counseling when needed, free support programs include the following:

Pet grief support groups: Some local animal shelters, humane societies, and veterinary schools offer pet grief support groups. If not, they may be able to refer you elsewhere. Call around until you find one.

Online support groups: If you can't find a local group near your home, there are groups that meet online or by phone. For example, chat rooms are hosted by the Association for Pet Loss and Bereavement *(www.aplb.org)*, Petloss.com *(petloss.com)*, and the Grief Support Center *(www.rainbowsbridge.com/Grief_Support_Center/Grief_Support_Home.htm)*.

Phone counseling: Several organizations offer pet loss support hotlines that anyone can call, including:

- ASPCA: (877) 474-3310

- Cornell University College of Veterinary Medicine: (607) 253-3932

- Tufts University School of Veterinary Medicine: (508) 839-7966

- Virginia-Maryland College of Veterinary Medicine: (540) 231-8038

- Washington State University College of Veterinary Medicine: (866) 266-8635

Open your heart to another animal in need: It may seem too difficult to open your heart to another animal, or too early, but consider it. Bringing a new pet into your home doesn't mean that you have given up on your search for your lost dog, are being disloyal, or have stopped caring about or mourning your missing dog. If you are not ready for a permanent commitment to a new dog, consider fostering a dog through a local rescue group or shelter on a temporary basis, to see how it feels to have a new dog in your life as you still grieve your lost dog.

Acknowledgements

This is a project I have thought about for a long time, and many people helped it become a reality. I am deeply grateful to the dog owners who shared their personal stories with me, some of whom I contacted out of the blue through a Facebook message: Stephanie Felicies, Jon Luke, Andrew Newman—whose dog Scout lit the spark of curiosity that led to this book—Kery O'Bryan, Kathleen Best and the late Andrew Schneider, Randi Spivak and Andy Kerr, Amber Yaw, and Ann Yoder.

I extend my heartfelt thanks to the scientists and canine experts who took the time to speak with me about their important work: Kat Albrecht, Dr. Marc Bekoff, Chris Boyer, Elizabeth Bradley, Landa Coldiron, Bill Dotson, Mary Helinski, Dr. Anneke Lisberg, Zita Macinanti, and Dr. Emily Weiss.

My deep appreciation goes to the professionals who patiently helped turn my manuscript into a real book:

editors Adrienne Hovey, Julie Miller, and Heather Harris; designer Lindsay Peternell; and the multi-talented Jane Ryder.

I am indebted to all those who generously provided feedback on my manuscript: Jessica Ennis, Paola Ferraris, Adine Forman, Elyse Forman, Liz Heyd, Jill Kind, Jon Luke, Una Song, Lindsey Stratton, and Larry Woodward. A special thanks to my fellow writers at the Petworth Library Writers Workshop, whose invaluable insights directly led to great improvements. I knew it would be helpful, but I had no idea how inspiring it would be to get feedback from other writers. And a special shout out to Julie Teel Simmonds and Melanie Sloan, whose beloved dogs inadvertently helped shape the book. Thank goodness for happy endings.

Online Resources

1. Tools and tips for effective lost dog posters and flyers:

www.helpfindlostpets.com/lost-dog-recovery-guide

www.petfbi.org/lost-pet-flyer.aspx

www.petbond.com/flyerentry.php

www.missingpetpartnership.org/recovery-tips/posters-5555/

www.lostdogsearch.com/fliersandsigns.htm

www.missingpetpartnership.org/recovery-tips/how-to-tag-your-car/

2. Free online databases to post missing dogs:

www.petfbi.org

www.flealess.org/lostpets/

www.fidofinder.com

missingpets.com

www.thecenterforlostpets.com

www.petharbor.com

www.helpinglostpets.com

3. Paid services to publicize missing dogs:

www.findtoto.com (phone alerts)

www.petamberalert.com (phone, fax, and social-media services)

www.lostmydoggie.com (phone, fax, e-mail, and mail services)

www.lostpetcards.com (postcard services)

www.usps.com/business/every-door-direct-mail.htm (direct mail)

Notes

Introduction

4 ***worried owners:*** Many people with dogs prefer to be called a dog "guardian," "parent," "companion," or some other term. For the purposes of clarity, this book uses the term "owner." I have also strived, throughout the book, to use "who" instead of "that," and "they" instead of "it" when referring to animals, in accord with the Animals and Media guidelines found at: www.animalsandmedia.org/.

5 ***almost half of American households:*** American Pet Products Association, "2015-2016 APPA National Pet Owners Survey Statistics: Pet Ownership & Annual Expenses." www.americanpetproducts.org/press_industrytrends.asp.

5 ***14 percent of dogs were lost:*** Emily Weiss, Margaret Slater, and Linda Lord, "Frequency of Lost Dogs and Cats in the United States and the Methods

Used to Locate Them," Animals 2 (2012): 301-315, doi: 10.3390/ani2020301.

5 *"One out of every three pets will go missing at some point in their lifetime":* American Kennel Club, "An Owner's Manual to: What to do if your pet goes missing," 12.

Chapter 1: The Dog-Human Bond

8 *Dogs and modern wolves share 99.96 percent of their DNA:* John Bradshaw, *Dog Sense* (New York: Basic Books, 2011), 1.

8 *Wolves and humans also had their differences:* Mark Derr, *How the Dog Became the Dog* (New York: The Overlook Press, 2011). The evolutionary and symbiotic relationship between humans and wolves is discussed at length in this book.

9 *share approximately 99 percent of our DNA:* Ann Gibbons, "Bonobos Join Chimps as Closest Human Relatives," *Science,* June 13, 2012, www.sciencemag.org/news/2012/06/bonobos-join-chimps-closest-human-relatives.

10 *higher levels of the chemical oxytocin:* Miho Nagasawa et al, "Oxytocin-gaze positive loop and the coevolution of human-dog bonds," *Science,* 17 (April 2015): 333-336, doi: 10.1126/science.1261022.

10 *researchers used MRI scanners:* Gregory Berns, How Dogs Love Us (New York: New Harvest, 2013) 203-204.

10 *The researchers concluded that:* Márta Gácsi et al, "Attachment behavior of adult dogs (Canis familiaris) living at rescue centers," *Journal of Comparative Psychology,* 115 (Dec 2001): 423-431.

11 *Their stress was lowest when they were around humans:* T. F. Pettijohn et al, "Alleviation of separation distress in 3 breeds of young dogs," *Developmental Psychobiology,* 10 (1977): 373-81. doi: 10.1002/dev.420100413.

11 *both young and adult dogs feel more secure:* Lisa Horn et al, "The Importance of the Secure Base Effect for Domestic Dogs—Evidence from a Manipulative Problem-Solving Task," *PLOS ONE* 8 (May, 2013). doi: 10.1371/journal.pone.0065296.

11 *they don't experience the oxytocin positive-feedback loop:* Nagasawa et al, "Oxytocin-gaze positive loop."

12 *a human family isn't the same:* Marc Bekoff, telephone interview with the author, September 28, 2015. Unless otherwise noted, all quotations attributed to Marc Bekoff are from this interview.

12 *"in the moment, there is something else happening":* Anneke Lisberg, telephone interview with the author, September 9, 2016. Unless otherwise noted, all quotations attributed to Anneke Lisberg are from this interview.

Chapter 2: Is My Dog Like Lassie?

13   *some over distances of hundreds:* Brad Steiger and Sherry Hansen Steiger, *Four-Legged Miracles* (New York: St. Martin's Griffin, 2013).

20   *Ann Yoder told me the story:* Ann Yoder, telephone interview with the author, May 24, 2017. All quotations attributed to Ann Yoder are from this interview.

Chapter 3: How Wolves Navigate

22   *the smartest predator in the wildlife kingdom:* Vilmos Csányi, *If Dogs Could Talk* (English translation) (New York: North Point Press, 2005) 10.

22   *Wolf packs generally consist of:* "Wolf FAQs," International Wolf Center, www.wolf.org/wolf-info/basic-wolf-info/wolf-faqs/#f.

23   *The bonds between each pack member are essential:* "The Bond," Living with Wolves, www.livingwithwolves.org/about-wolves/general-infomration/.

23   *wolves can roam 50 miles or more a day:* "Wolf FAQs," International Wolf Center, http://www.wolf.org/learn/basic-wolf-info/wolf-faqs/.

24   *they also invent new shortcuts and detours:* Roger Peters, *Dance of the Wolves,* (New York: McGraw-Hill Book Company, 1984), 199.

24   *what are now known as "mental maps":* Peters, *Dance of the Wolves,* 219-221.

Notes

25 **88 wolves in the western United States:** Elizabeth Bradley et al, "Evaluating Wolf Translocation as a Nonlethal Method to Reduce Livestock Conflicts in the Northwestern United States," *Conservation* Biology (2005): 1498–1508. doi: 10.1111/j.1523-1739.2005.00102.x.

25 **There were only four translocation events:** Elizabeth Bradley, e-mail message to author, April 14, 2017.

Chapter 4: The Science of Dogs and Navigation

27 **Ain't no mountain high enough:** Nickolas Ashford and Valerie Simpson, "Ain't No Mountain High Enough."

27 **the domestic dog's closest living relative:** "Ancient Wolf Study Points to Early Split Between Wolf, Dog Lineages," GenomeWeb, last modified May 21, 2015, www.genomeweb.com/sequencing-technology/ancient-wolf-study-points-early-split-between-wolf-dog-lineages.

28 **When the fence was U-shaped:** Harry Frank and Martha Gialdini Frank, "Comparison of Problem-Solving Performance in Six-week-old Wolves and Dogs," *Animal Behavior* 30 (1982): 95-98.

29 **the wolf pups were both more perseverant:** Harry Frank and Martha Gialdini Frank, "Comparative Manipulation-Test Performance in Ten-Week-Old Wolves (Canis lupus)and Alaskan Malamutes (Canis familiaris): A Piagetian Interpretation," *Journal of Comparative Psychology* 99 (1985): 266-274.

30   *while dogs can use landmarks:* Brian Hare and Vanessa Woods, *The Genius of Dogs* (New York: Dutton, 2013), 154.

Chapter 5: Scout

33   *she was extremely smart:* Andrew Newman, telephone interview with the author, August 5, 2015. All quotations attributed to Andrew Newman are from this interview.

37   ***It was the moment Andrew had dreamed about:*** "Boulder Man Finds Lost Dog After 4 Weeks," The Denver Channel, July 30, 2009, www.thedenverchannel.com/lifestyle/family/boulder-man-finds-lost-dog-after-4-weeks.

Chapter 6: Why Dogs Go Missing

41   *mark those spots with his own urine:* Elizabeth Marshall Thomas, *The Hidden Life of Dogs* (Boston: First Mariner Books, 1993), 24-25.

41   *dogs have "massive" olfactory bulbs:* Berns, *How Dogs Love Us,* 196-197.

42   *something "novel and unexpected":* Temple Grandin and Catherine Johnson, *Animals in Translation* (New York: Scribner, 2005), 44-48.

42   *An animal in a state of panic:* Grandin and Johnson, *Animals in Translation,* 190.

42   *the most common reason that dogs go missing:* Zita Macinanti, telephone interview with the author, February 9, 2016. All quotations attributed to Zita Macinanti are from this interview.

Notes

44   ***people who find stray dogs:*** Emily Weiss, telephone interview with the author, September 20, 2016.

Chapter 7: Stolen Dogs

45   ***Reports of pet theft:*** Brad, Tuttle, "'Pet Flipping' Is Now a Thing," *Time Magazine,* July 16, 2013, http://business.time.com/2013/07/16/pet-flipping-is-now-a-thing/.

45   ***being used in dogfighting rings:*** Royale Da, "Dogs being swiped for possible fighting ring, deputies say," KOAT-TV, March 5, 2015.

46   ***little evidence that pets are stolen for research:*** National Research Council of the National Academies, "Scientific and Humane Issues in the Use of Random Source Dogs and Cats in Research," The National Academies Press, 2009, 84.

46   ***dogs have been stolen by animal abusers:*** Howard Pankratz, "Man who dragged dog to death gets maximum sentence," *The Denver Post,* July 30, 2010, updated May 5, 2016, www.denverpost.com/2010/07/30/man-who-dragged-dog-to-death-gets-maximum-sentence/.

46   ***a yellow Labrador retriever puppy:*** Andrew Schneider, telephone interview with the author, January 22, 2016. All quotations attributed to Andrew Schneider are from this interview.

49   ***a five-month-old pit bull puppy:*** Stephanie Felicies, telephone interview with the author,

February 16, 2017. All quotations attributed to Stephanie Felicies are from this interview.

51   ***A TV news crew:*** "Parkchester woman offers reward for stolen puppy," News 12, bronx.news12.com/story/34803473/parkchester-woman-offers-reward-for-stolen-puppy.

Chapter 8: Dora

53   ***a stray black and tan German shepherd mix:*** Kery O'Bryan, telephone interview with the author, July 11, 2015. All quotations attributed to Kery O'Bryan are from this interview.

56   ***An animal control officer had trapped:*** Collin County Public Information Office, "This Dog's Tale," http://www.co.collin.tx.us/public_information/features/Pages/dogs_tale.aspx.

56   ***There's a video:*** Collin County Animal Services, "Lost Dog Dora reunited with family after seven months!," www.youtube.com/watch?v=120gUXl7iXg.

Chapter 9: Top Ten Tips to Keep Your Dog Safe at Home

61   ***a 200% higher chance of being returned:*** "An Owner's Manual to: What to do if your pet goes missing," American Kennel Club, 15.

61   ***the Goldston family was contacted:*** Jonathan Gonzalez, "Dog missing for 9 years reunited with owners," 9 News, August 8, 2015,

Notes

www.9news.com/story/news/local/2015/08/08/missing-dog-foothills-animal-shelter/31351353/.

61 *Joshua Edwards was contacted:* Carli Teproff, "After 8 years, Coconut Grove dog owner reunites with his Rottweiler," *Miami Herald,* May 19, 2015, http://www.miamiherald.com/news/local/community/miami-dade/article21429372.html.

62 *someone who has scanned your dog's microchip:* "The Microchip Mess," *SwayLove.org,* July 17, 2013, www.swaylove.org/the-microchip-mess.

63 *patented anti-migration feature:* "FAQs," Home-Again, www.homeagain.com.

65 *"remained nervous and distressed":* Frank and Gialdini Frank, "Comparative Manipulation-Test Performance," 266-274.

69 *only 33 percent:* "Do Your Pets Wear an ID tag? They Should," American Society for the Prevention of Cruelty to Animals, Press release, September 21, 2011.

Chapter 10: Top Ten Tips to Find Your Missing Dog

75 *One study of lost dogs:* Weiss, Slater, and Lord, "Frequency of Lost Dogs and Cats in the United States and the Methods Used to Locate Them," 301-302.

76 *are found in industrial areas:* Landa Coldiron, telephone interview with the author, January 15, 2016. Unless otherwise noted, all quotations attributed to Landa Coldiron are from this interview.

78 ***search on the internet for a photo:*** Annalisa Berns and Landa Coldiron, *Lost Dog Recovery Guide,* self-published by International Pet Detectives, LLC, 2011, 24.

80 ***Amber adopted in 2010:*** Amber Yaw, telephone interview with the author, August 19, 2016.

82 ***I don't think we would've found Lucy:*** Kelli Lageson, "Internet key in finding lost dog," *Albert Lea Tribune,* June 2, 2011, www.albertleatribune.com/2011/06/internet-key-in-finding-lost-dog/.

84 ***banners, commercial-type signs:*** Berns and Coldiron, *Lost Dog Recovery Guide,* 53-66.

84 ***a lost toy poodle:*** Janet Elder, *Huck* (New York: Broadway Paperbacks, 2010).

88 ***hold dogs for only 72 hours:*** "Lost and Found," Animal Care Centers of New York City, nycacc.org/LostFound.htm.

88 ***Augustus, a stray dog:*** a video of August, the dog rescued by Heart of Alabama, Save. Rescue. Adopt, can be seen posted by AL.com at: https://www.youtube.com/watch?v=Laq8xTOyxoE.

90 ***"You are looking for a person":*** Kat Albrecht, telephone interview with the author, March 30, 2016. Unless otherwise noted, all quoted attributed to Kat Albrecht are from this interview.

91 ***Rain was always friendly and playful:*** Andy Kerr and Randi Spivak, interview with the author,

Notes

January 31, 2016. Unless otherwise noted, all quoted attributed to Andy Kerr and Randi Spivak are from this interview.

96 *a news investigation into one company:* Deanna Dewberry, "Lost Pet-Search Services: Helpful or Waste of Money?," NBCDFW, November 5, 2012, www.nbcdfw.com/investigations/Lost-Pet-Search-Services-Helpful-or-Waste-of-Money-177401881.html.

99 *His family went door to door:* Lisa Rathke, "Bacon, pet psychic turn up zilch in search for dog," *Associated Press,* December 10, 2014, www.seattletimes.com/life/pets/bacon-pet-psychic-turn-up-zilch-in-search-for-dog/.

99 *more than a year and a half:* Rose Spillman, "Missing dog brought home after 18 months on the run," WCAX, January 10, 2016, www.wcax.com/story/30928020/missing-dog-brought-home-after-18-months-on-the-run. See also: Mitch Wertlieb and Kathleen Masterson, "The Odyssey Of Murphy The Dog, And A Reporter Who Became Part Of The Story," January 15, 2016, digital.vpr.net/post/odyssey-murphy-dog-and-reporter-who-became-part-story.

101 *SAR dogs detect scents:* Chris Boyer, phone interview with the author, January 14, 2016. Unless otherwise noted, all quoted attributed to Chris Boyer are from this interview.

103   *very few lost dogs are actually found:* Danielle Robertson, "When NOT to Use a Tracking Dog to Find a Lost Dog," Lost Pet Research Blog, January 20, 2015, lostpetresearch.com/2015/01/not-to-use-a-tracking-dog-to-find-a-lost-dog/.

103   *a qualified pet-finder dog would have to be trained:* Bill Dotson, interview with the author, January 15, 2016. Unless otherwise noted, all quoted attributed to Bill Dotson are from this interview.

103   *the key to the success:* Mary Helinski, interview with the author, March 3, 2015.

105   *your dog is somewhere:* Kat Albrecht, *The Lost Pet Chronicles* (New York: Bloomsbury Publishing, 2004), 233.

Epilogue

106   *grieving a missing pet can be harder:* "Pet Loss: When a Pet Goes Missing," *Grief Healing Blog*, http://www.griefhealingblog.com/2014/10/pet-loss-when-pet-goes-missing.html.

106   *his Newfoundland Ursula:* Jon Luke, interview with the author, May 25, 2017. Unless otherwise noted, all quotations attributed to Jon Luke are from this interview.

www.ingramcontent.com/pod-product-compliance
Lightning Source LLC
Chambersburg PA
CBHW032043290426
44110CB00012B/932